THE FUTURE
OF GENETICS

GENETICS & EVOLUTION

THE FUTURE OF GENETICS

Beyond the Human Genome Project

RUSS HODGE

FOREWORD BY NADIA ROSENTHAL, PH.D.

Facts On File

An imprint of Infobase Publishing

This book is dedicated to the memory of my grandparents, E. J. and Mabel Evens and Irene Hodge; to my parents, Ed and Jo Hodge; and especially to my wife, Gabi, and my children, Jesper, Sharon, and Lisa, with love.

THE FUTURE OF GENETICS: Beyond the Human Genome Project

Copyright © 2010 by Russ Hodge

Facts On File, Inc.
An imprint of Infobase Publishing
132 West 31st Street
New York NY 10001

Library of Congress Cataloging-in-Publication Data

Hodge, Russ.
   The future of genetics : beyond the human genome project / Russ Hodge ; foreword by Nadia Rosenthal.
      p. ; cm. — (Genetics and evolution)
   Includes bibliographical references and index.
   ISBN 978-0-8160-6684-1 (alk. paper)
   1. Genetics—Popular works. 2. Genetics—Forecasting. I. Title. II. Series: Genetics and evolution.
   [DNLM: 1. Genetics, Medical. 2. Genome, Human. 3. Molecular Biology—methods. QZ 50 H688f 2010]

   QH437.H63 2010
   303.48'3—dc22   2009018297

Text design by Kerry Casey
Illustrations by Sholto Ainslie
Photo research by Elizabeth H. Oakes
Composition by Hermitage Publishing Services
Cover printed by Bang Printing, Brainerd, MN
Book printed and bound by Bang Printing, Brainerd, MN
Date printed: March 2010
Printed in the United States of America

This book is printed on acid-free paper.

I say that it touches a man that his blood is sea water and his tears are salt, that the seed of his loins is scarcely different from the same cells in a seaweed, and that of stuff like his bones coral is made. I say that the physical and biologic law lies down with him, and wakes when a child stirs in the womb, and that the sap in a tree, uprushing in the spring, and the smell of the loam, where the bacteria bestir themselves in darkness, and the path of the sun in the heaven, these are facts of first importance to his mental conclusions, and that a man who goes in no consciousness of them is a drifter and a dreamer, without a home or any contact with reality.

—*An Almanac for Moderns: A Daybook of Nature,*
Donald Culross Peattie, copyright ©1935
(renewed 1963) by Donald Culross Peattie

Contents

Foreword

Science has played an increasingly central role in human affairs over the last two centuries, affecting every aspect of our daily lives. From the time of the Industrial Revolution, we have somewhat taken for granted our increasing dependence on the products of mechanical ingenuity that drive agriculture, manufacturing, mining, transportation, and communications. At the beginning of the 19th century, our great-grandparents could not easily have imagined our modern, engine-powered world. When we try to conceive of the advances that our great-grandchildren will witness, it is likely to be the fruits of genetic research that will change lives most dramatically.

The Future of Genetics offers an exciting, sometimes startling survey of the mechanics of today's genetic research at the dawn of a new century and provides a glimpse into its future. In chapter 1, Russ Hodge shows how the last century of genetic research has paved the way for the rapid progress we are witnessing in the field. Increasingly sophisticated tools of genetic analysis and manipulation are currently being developed in laboratories across the world. Drawing from other research streams, such as advances in molecular and cell biology, imaging, and informatics, today's genetics is offering a view of nature that is richer and more varied than we could have imagined previously.

The new tools of genetics also suggest prospects for changing the course of natural selection, sometimes in ways that are every bit as bizarre as science fiction. In chapter 2, Hodge explores how current advances in genetics have infiltrated and influenced our societal views and values. In a world where the communication of information is not only instantaneous but can also be manipulated to suit the needs of politicians, journalists, and the marketplace, a little information can be a dangerous thing.

In these accounts, geneticist are often morphed into the monsters they are supposedly creating, and the relevance of their extraordinary discoveries is sometimes lost on a public craving sensationalism.

Despite the view that putting the power of genetics in the hands of amoral inventors is hazardous, human curiosity is insatiable, and the outcome of genetic research is already beginning to make a major difference in our lives. Whether in regard to the early detection and prevention of congenital disease, the promise of a longer and healthier life span through a better understanding of the role of genes in aging, or the creation of new drugs based on a person's unique genetic makeup, the impact of genetics on medicine will continue to grow. Chapter 3 reviews the progress that has already been made in harnessing new knowledge gleaned from genetics to control our personal destinies and looks forward to the ways we may handle these changes as a society.

Other areas of human society, as explored in chapter 4, are just as affected by the new genetics as they were by the mechanics of the Industrial Revolution. Experimentation with genetically modified crops has not been a popular aspect of this progress, but in times of food shortages and global climate change, genetically modified crops are already affecting agriculture in our lifetime. Other directions that genetic research may take may be just as inevitable, but their cost might be too great: The creation of new designer organisms or the resuscitation of extinct species, for example, might be exciting grist for science fiction writers but could have disastrous consequences similar to the ill-advised introduction of cane toads into Australia in the 1930s for agricultural pest control that has resulted in a plague. The more genetics teaches about biodiversity, the more we appreciate how our own population explosion and the environmental devastation we create in the name of sustaining our way of life threatens Earth's ecosystem, upon which we vitally depend. Just as we have evolved from the days of coal-driven engines by developing new, quieter, and less polluting versions, the mechanics of tomorrow will need to harness the power of genetics for the benefit of society and the protection of the environment. This is a responsibility we cannot afford to ignore.

—Nadia Rosenthal, Ph.D.
Head of Outstation
European Molecular Biology Laboratory
Rome, Italy

Preface

In laboratories, clinics, and companies around the world, an amazing revolution is taking place in our understanding of life. It will dramatically change the way medicine is practiced and have other effects on nearly everyone alive today. This revolution makes the news nearly every day, but the headlines often seem mysterious and scary. Discoveries are being made at such a dizzying pace that even scientists, let alone the public, can barely keep up.

The six-volume Genetics and Evolution set aims to explain what is happening in biological research and put things into perspective for high-school students and the general public. The themes are the main fields of current research devoted to four volumes: *Evolution, The Molecules of Life, Genetic Engineering,* and *Developmental Biology.* A fifth volume is devoted to *Human Genetics,* and the sixth, *The Future of Genetics,* takes a look at how these sciences are likely to shape science and society in the future. The books aim to fill an important need by connecting the history of scientific ideas and methods to their impact on today's research. *Evolution,* for example, begins by explaining why a new theory of life was necessary in the 19th century. It goes on to show how the theory is helping create new animal models of human diseases and is shedding light on the genomes of humans, other animals, and plants.

Most of what is happening in the life sciences today can be traced back to a series of discoveries made in the mid-19th century. Evolution, cell biology, heredity, chemistry, embryology, and modern medicine were born during that era. At first these fields approached life from different points of view, using different methods. But they have steadily grown closer, and today they are all coming together in a view of life that stretches from single molecules to whole organisms, complex interactions between species, and the environment.

The meeting point of these traditions is the cell. Over the last 50 years biochemists have learned how DNA, RNA, and proteins carry out a complex dialogue with the environment to manage the cell's daily business and to build complex organisms. Medicine is also focusing on cells: Bacteria and viruses cause damage by invading cells and disrupting what is going on inside. Other diseases—such as cancer or Alzheimer's disease—arise from inherent defects in cells that we may soon learn to repair.

This is a change in orientation. Modern medicine arose when scientists learned to fight some of the worst infectious diseases with vaccines and drugs. This strategy has not worked with AIDS, malaria, and a range of other diseases because of their complexity and the way they infiltrate processes in cells. Curing such infectious diseases, cancer, and the health problems that arise from defective genes will require a new type of medicine based on a thorough understanding of how cells work and the development of new methods to manipulate what happens inside them.

Today's research is painting a picture of life that is much richer and more complex than anyone imagined just a few decades ago. Modern science has given us new insights into human nature that bring along a great many questions and many new responsibilities. Discoveries are being made at an amazing pace, but they usually concern tiny details of biochemistry or the functions of networks of molecules within cells that are hard to explain in headlines or short newspaper articles. So the communication gap between the worlds of research, schools, and the public is widening at the worst possible time. In the near future young people will be called on to make decisions—large political ones and very personal ones—about how science is practiced and how its findings are applied. Should there be limits on research into stem cells or other types of human cells? What kinds of diagnostic tests should be performed on embryos or children? How should information about a person's genes be used? How can privacy be protected in an age when everyone carries a readout of his or her personal genome on a memory card? These questions will be difficult to answer, and decisions should not be made without a good understanding of the issues.

I was largely unaware of this amazing scientific revolution until 12 years ago, when I was hired to create a public information office at one of the world's most renowned research laboratories. Since that time I have had the great privilege of working alongside some of today's greatest researchers, talking to them daily, writing about their work, and picking their brains about the world that today's science is creating. These books aim to share those experiences with the young people who will shape tomorrow's science and live in the world that it makes possible.

Acknowledgments

This book would not have been possible without the help of many people. First I want to thank the dozens of scientists with whom I have worked over the past 12 years, who have spent a great amount of time introducing me to the world of molecular biology. In particular I thank Volker Wiersdorff, Patricia Kahn, Eric Karsenti, Thomas Graf, Nadia Rosenthal, Walter Rosenthal, and Walter Birchmeier. My agent, Jodie Rhodes, was instrumental in planning and launching the project. Frank K. Darmstadt, executive editor, kept things on track and made great contributions to the quality of the text. Sincere thanks go as well to the production and art departments for their invaluable contributions. I am very grateful to Beth Oakes for locating the photographs for the entire set. Finally, I thank my family for all their support. That begins with my parents, Ed and Jo Hodge, who somehow figured out how to raise a young writer, and extends to my wife and children, Gabi, Jesper, Sharon, and Lisa, who are still learning how to live with one.

Introduction

The Future of Genetics was inspired by a few facts and experiences that most people have shared in one way or another. My grandfather was born in the year 1900 and died toward the end of the 20th century, having witnessed such amazing transformations that the world must have seemed almost alien. For example, discoveries had led to airplanes, televisions, computers, and the Internet, to name only a few. Many of the changes were comforting to a middle-class family in the developed world: electricity everywhere, all kinds of luxuries to make daily life easier, vast amounts of information and entertainment transmitted directly into the home, and great advances in medicine, including the insulin that extended his life.

However, these positive changes came at a pace that was difficult to adjust to, with some worrisome side effects. The hunger for energy led to widespread strip-mining of coal in his home state, and burning the coal caused smog and global warming. Modern physics had produced weapons capable of wiping out human civilization and nuclear power plants that dotted the countryside—were they really safe? Industrial chemicals and other types of pollution ate at the ozone layer and contaminated the soil. The technology that permitted the creation of cell phones and the Internet could be used to eavesdrop on people. And the same methods that produced insulin were being used to insert new genes into crops grown by his neighbors, with consequences that might bring risks; it was hard to be sure.

Many people seem to feel that scientific discoveries and progress are inevitably accompanied by unforeseen dangers. This can be witnessed anywhere. Start a conversation about genetic engineering at any café, classroom, or dinner party. Phrases such as "scientists should not meddle with nature," or "the Earth will take revenge on humans," or "as doctors get smarter, nature will invent smarter diseases" are bound to be heard.

Where do these beliefs come from? Are they a valid appreciation of real risks, an impression given by highly exaggerated scenarios of science fiction, have people's feelings about the biological sciences been unfairly contaminated by experiences with other types of research and technology? Is cloning really anything to be scared of or have people simply seen too many bad science fiction films? Is biological research itself dangerous or do problems arise from the pressure to turn science into products that can be catapulted onto global markets? Are people afraid of science or are they afraid of change in general? This book aims to give readers the facts they need to find sensible answers to these questions and to distinguish serious issues from hype and unrealistic fears. That is difficult at a time when the line between information, entertainment, and ideologies has become blurred. Optimistic press releases are released by companies trying to find investors; science fiction films sensationalize, exaggerate, and dramatize rather than explain, and scientific data is being misused by people with political or religious agendas.

It is important to try to form a realistic picture of how science might change the future, even though speculations about the distant future of technology often turn out to be wrong—where are the starships or time machines imagined just a few decades ago? On the other hand, some visions do come true. Today's biology is having an enormous impact on medicine, particularly in diagnosing disease and understanding how drugs work, and researchers strongly believe that it will soon change the way doctors cope with complex illnesses like cancer. It may even provide treatments for genetic diseases. Many of the possibilities for therapies or technologies foreseen by scientists already exist in some form. The question is whether they can be extended to humans safely and in a controlled way. Change may come more rapidly than anyone anticipates, and there may be intense pressure to sell technologies before they have been adequately tested. This can already be seen in the case of some drugs. Some medications for attention-deficit hyperactivity disorder (ADHD) were prescribed to children before their long-term effects on children's development were fully understood.

The ethical issues need to be confronted now, in preparation for a future in which biotechnology is more powerful. Yet making intelligent decisions will require a realistic understanding of how science works. Research has become thoroughly global, which means that the only way to regulate certain practices may be for the global research community as a whole to agree that some things should not be done.

Currently, there are areas of widespread agreement. Few scientists see the sense in cloning a person—which means taking material from a person's cells and producing an identical twin at a later date. On the other hand, it would be useful to know how to stimulate embryonic stem cells into rebuilding damaged nerves, muscles, or heart tissue—and accomplishing one of these things might require learning to do the other.

Most scientists also agree that a person's genes should not be altered in a way that can be passed along to his or her children. But just a century ago, many scientists—and others—felt differently. They promoted misguided efforts to improve humanity in the first half of the 20th century in programs that forcibly sterilized thousands of mental patients and other "undesirables" throughout the United States. This thinking culminated in the Holocaust in Nazi Germany, whose perpetrators also claimed to be trying to improve mankind. Ever since, society has rejected the idea of tampering with the human genome. But people might find it hard to turn their backs on the possibility of ridding their own families of a mutation that causes the symptoms and suffering of Huntington's or Alzheimer's, cancer, or some other devastating disease.

The Future of Genetics considers where research in genetics, molecular biology, and medicine is headed while trying to cleanly separate facts from fiction and ideologies. The first chapter sets the stage by showing how over the last 150 years different strands of biological research have become interwoven to create a new kind of interdisciplinary science. These trends have been accompanied by works of fiction—from *Frankenstein* to *Brave New World* to *Jurassic Park*—in which authors have explored the social impacts and ethical implications of discoveries. Chapter 2 explores how some of these famous works—and other science-related events—have shaped people's perceptions of science. Chapter 3 presents a range of very new technologies that are giving scientists a broader view of life and providing new ways to manipulate organisms and the environment. The final chapter focuses on some of the most fascinating questions that scientists are posing about the future: the causes of aging and death, the nature of the brain and mind, and the future of life on Earth. Genetics is playing a key role in research into all of these areas; it may also be the gateway to improving people's lives and ensuring that the Earth remains a hospitable place to live.

1

The Origins of Twenty-first-Century Biology

The 21st century opened with the completion of a working draft of the human *genome sequence.* After more than 15 years of intense effort that involved thousands of people working day and night around the globe, researchers finally finished cracking the entire human genetic code. The text consists of only four letters—the chemical subunits called *nucleotides,* or bases, that make up *deoxyribose nucleic acid,* or *DNA.* When 3 billion of such subunits are strung together, they contain enough information to build a person. Just 50 years after the discovery that genes were made of DNA, the code was deciphered. At the moment, the state of that information is more like the contents of a kitchen's shelves than a recipe book. The current challenge is to learn how cells use the ingredients to create human beings and every other form of life on Earth.

The human genome encodes a vast amount of information about human evolution and the way a fertilized egg cell develops into a body. It also holds clues to the origins of diseases such as cancer and the mechanisms behind processes like aging. To make use of this information, scientists will have to learn how the genome produces other types of molecules—*RNAs* and *proteins*—and discover how they control the structure and behavior of cells.

Today's biology is a mix of methods and concepts from fields such as chemistry, physics, genetics, embryology, evolution, mathematics, and medicine. Interestingly, nearly all of these modern fields can be traced back to a series of breakthroughs that occurred within a few decades of the mid-19th century. The best way to understand today's science and its implications for the future is to take a brief look backward and see how these disciplines have become interwoven in the intervening years.

PLANTS, ANIMALS, AND CELLS

In 1833, Professor Johannes Müller (1801–58) moved from the city of Bonn, Germany, to take up a new position at the Humboldt University of Berlin. Until his death 25 years later, he carried out research into human senses and the nervous system while training an entire generation of young scientists. In the early 19th century, Germany, England, and a few other European countries were the world's hotspots of scientific discovery. Several of Müller's students went on to revolutionize—and in some cases create—the modern fields of cell biology, embryology, and medicine.

When Müller moved to Berlin, he brought along a talented student named Theodor Schwann (1810–82). One day at the Berlin train station, Schwann struck up a conversation with Mathias Schleiden (1804–81), a fellow student. Schleiden had studied law in Heidelberg in southern Germany, then moved north to Hamburg to open a law practice, but there he suffered bouts of depression that led to an unsuccessful attempt at suicide. As a result, he carried a bullet in his brain for the rest of his life. Now he had moved to Berlin to start over, this time as a botanist. He too was taking classes with Müller.

Their professor had become an enthusiastic user of a new type of microscope that had just been invented by the Englishman Joseph Jackson Lister. Up to that time microscopes had been limited by problems of blurring and distortion. Lister's innovation was to mount two lenses at fixed distances from each other in a tube, a construction that dramatically increased

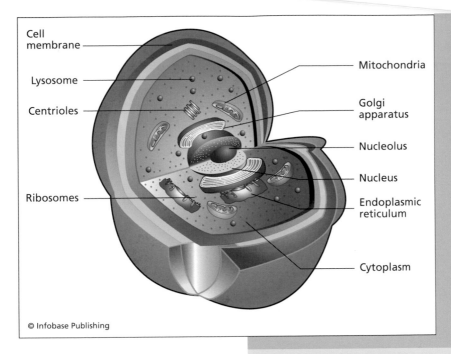

Cell membrane

Lysosome

Centrioles

Ribosomes

Mitochondria

Golgi apparatus

Nucleolus

Nucleus

Endoplasmic reticulum

Cytoplasm

© Infobase Publishing

Theodor Schwann and Mathias Schleiden played a key role in the birth of modern biology with their discovery that both plants and animals were made up of cells. New staining techniques soon began to reveal "organelles" and other substructures within the cells.

resolution and gave scientists their sharpest view ever of the microscopic world. Over the course of hundreds of hours studying plant specimens on carefully prepared slides, Schleiden discovered that plants were built entirely of fundamental units—single cells—which somehow formed from the nuclei of other cells.

One night at dinner with Schwann, Schleiden mentioned what he had found. Schwann was less interested in plants, but Schleiden's idea might explain what he had been seeing in animal tissue. The two men left their meal half-eaten and rushed over to Schwann's laboratory. Previously, anatomists had believed that animal tissues were made of fibers, grains, tubes, and other objects. They had been looking at cells, Schwann realized, without knowing it.

Discovering that bodies were made of more fundamental units was a huge leap for science. It gave researchers a new way to look at the formation of embryos, for example—as a collection of cells that developed in different ways to form various types of tissues. But neither man followed the idea to its logical conclusion. That would be the accomplishment of another of Müller's students, Rudolf Virchow (1821–1902), one of the greatest physicians of the 19th century.

Upon graduation, Virchow held a double professorship at the University of Berlin and the Charité hospital. There, he treated patients and carried out research related to fundamental questions about cells. In 1858, he took Schleiden and Schwann's observations a step further with his statement of the doctrine *Omnis cellula e cellula,* "Every cell originates from a similar, previously existing cell." Today, this is such a basic principle of biology that it seems obvious, but at the time many scientists believed that cells could somehow arise by themselves, in a process called *spontaneous generation,* crystallizing from fluids or more basic substances. Virchow's simple new idea had a huge impact because it changed the way scientists thought about all sorts of questions, from the growth of embryos to the nature of disease.

For example, it gave Virchow a new view of cancer. He realized that tumors arose from small pools of cells that divided too often in the wrong places. Removing the source might stop the spread of the disease. He developed new laboratory methods to diagnose cancer and new surgical procedures to treat it. Some of his ideas

German physician Rudolf Virchow first proposed the theory that every cell arises from an existing cell, which had a tremendous impact on embryology, the study of cancer and infectious diseases, and the rest of the life sciences.

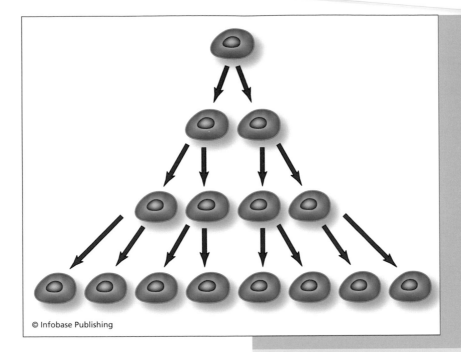

© Infobase Publishing

were ahead of their time—his experience with patients, for example, suggested that the disease often arose at sites of injuries or infections. This made him think that tumors might be linked to inflamma-tions—the body's response to injuries. For more than a century, most scientists rejected this idea, but there now is compelling evidence that he was right. Some types of cancer have been linked to infections by viruses, inflammations, and autoimmune diseases in which the body has trouble distinguishing between its own and foreign cells.

Omnis cellula e cellula—"Every cell originates from a similar, previously existing cell." This applies equally to tissues of the eye, a maggot that suddenly appears in rotting meat, or a tumor cell.

Virchow went on to become one of the most famous scientists in the world, a Renaissance man passionately interested in other sciences such as archaeology (he participated in the archaeologist Heinrich Schliemann's first excavations of Troy) and a bold social thinker. Thanks to his efforts, Berlin developed a modern sanitation system that greatly improved the health of the city. He was elected to parliament where he pushed for democratic reforms. This made him such a thorn in the side of Prime Minister Otto von Bismarck that the statesman challenged him to a duel. Virchow's response was to laugh.

CELL BIOLOGY AND MEDICINE

Virchow's discovery of the connection between cancer and cells helped found a new type of medicine. Within a few years, the transformation of medicine into a modern science would be complete with the discovery that bacteria—also cells—were responsible for a wide range of epidemics such as cholera, tuberculosis, and the plague. Two major figures in this revolution were the German physician Robert Koch (1843–1910) and the French scientist Louis Pasteur (1822–95).

The idea that diseases were caused by tiny parasites had a historical precedent. In 36 B.C.E., the Roman scholar Marcus Terrentius Varro had warned people not to build their homes too close to swamps because such areas "breed certain minute creatures which cannot be seen by the eyes, but which float in the air and enter the body through the mouth and nose and cause serious diseases." Arabic physicians of the Middle Ages suggested that microscopic substances were responsible for infectious diseases. But well into the 19th century, most European physicians still held to a *miasma theory of disease,* which suggested that tiny particles of decomposed material floated through the air, accompanied by unpleasant smells, and caused sickness through poisoning. The idea was useful in a way: It encouraged improvements in sanitation that often resulted in better health. Clean air and water usually held far fewer dangerous microorganisms. Attributing disease to bad air, however, was like a sleight of hand in which a magician gets the public to look at the wrong hand.

The widening use of microscopes began to change this situation. In 1840, Friedrich Henle (1809–85), one of Johannes Müller's assistants in Berlin, wrote "On Miasma and Contagia," an essay in which he challenged the prevailing theory. No one had ever proven that miasma existed, he wrote; it was more likely that the air was merely the route taken by tiny living parasites as they moved from one host to another. Disease organisms had not been found, he said, because they looked so similar to the tissues they infected. While Henle had yet to identify such organisms, he was confident that they would be found, and he joined Müller in pushing medical students to spend time at the microscope.

As this new idea of disease was being introduced in Germany, a researcher in northern France was about to perform an important experiment that would help confirm it. Louis Pasteur was a gifted chemist who had become more and more interested in microorganisms. In the early 1850s, he took on the question of *fermentation*—the type of chemical transformation that occurred in the production of wine and other types of alcoholic drinks. People had made use of the phenomenon for thousands of years without understanding why it happened. Pasteur proved that this chemical problem was actually a biological one. Within fermenting liquids he discovered yeast cells. He wrote, "I am of the opinion that alcoholic fermentation never occurs without simultaneous organization, development and multiplication of cells."

Where did the cells come from? Virchow's theory of *Omnis cellula e cellula* was still new. Most researchers were convinced that complex, visible organisms such as maggots came from eggs already found in rotting meat or which had somehow drifted onto it, but the origins of microorganisms were less clear. In 1864, Pasteur published the results of a very careful set of experiments in which he demonstrated that microbes such as bacteria or yeast could not arise in sterile conditions. "Never will the doctrine of spontaneous generation recover from the mortal blow struck by this simple experiment," he wrote.

The next step was to develop a cellular view of disease. Henle had moved to the University of Göttingen in central Germany; one of his students there, Robert Koch, undertook the search for disease organisms in earnest. He began with the bacterium that caused anthrax, a serious disease that affected cows, sheep, and other grazing animals. Humans could catch it by eating meat from infected animals or being exposed to their fur or wool. The bacterium had been discovered in the blood of

This stamp honors the German physician Robert Koch, who was the first to prove a connection between a disease and a specific microbe—the bacterium responsible for anthrax. Koch established criteria that are still fundamental to identifying the causes of infectious diseases. (*Deutsche Reichspost*)

sheep in 1850 by a French researcher named Casimir Davaine, but its connection to the disease was not entirely clear; animals sometimes contracted anthrax without having had contact with other sick animals. Koch's experiments showed that in its normal form the organism could only survive a short time outside of a host, but it was also capable of forming capsulelike spores that could lie dormant on a field for long periods of time. Ingested by grazing animals, the microbes could become active again and trigger the disease. The work established the first definitive link between a bacterium and illness. Thinking that other disease organisms might also survive outside the body, Koch pushed hospitals in Berlin to begin sterilizing their surgical instruments. He went on to improve methods for growing microorganisms in cell cultures and staining them so that they could be seen more easily under the microscope.

Koch's next discovery was *Mycobacterium tuberculosis,* the cause of tuberculosis. Long one of mankind's worst diseases, it was responsible for one of every seven deaths in the mid-19th century. The finding was considered so important that Koch was awarded the 1905 Nobel Prize in physiology or medicine.

He devoted the next several years of his life to trying to develop a sort of vaccine made of extracts from the bacterium. Although it did not work and his career suffered as a result, Koch had a profound impact on the development of modern medicine. He developed a set of conditions, now known as Koch's postulates, which had to be fulfilled to prove that a particular microbe was responsible for a disease. The following principles are still considered fundamental to disease research:

- the microorganism has to be found in every patient or organism suffering from the disease
- researchers must be able to grow it in pure cultures in the laboratory
- even after several generations of growth in the laboratory, it must still be capable of causing the disease
- if it has been artificially introduced in an animal and causes the disease, researchers must be able to extract and culture it again

Armed with Koch's principles and methods, his students went on to find the microbes responsible for typhoid, leprosy, the bubonic plague, and other major diseases.

Louis Pasteur was having much better success in the search for cures. As he worked on a form of cholera that infected chickens, a series of chance events led him to some of the basic principles underlying vaccination. He was growing cholera bacteria in the laboratory and injecting it into the birds. One round of cultures became spoiled. When he tried to use it to infect a new round of chickens, they did not develop the disease. He tried again, using the same birds and a fresh batch of the microbe. They became sick but made a complete recovery. Pasteur reasoned that something about dead or weakened microbes could protect the birds—and possibly people—from future infections.

This phenomenon had been seen before. In the 18th century, physicians had begun inoculating people with bits of tissue taken from the sores of a victim of smallpox, often giving them a milder form of the disease that protected them later. In 1777, concerned that an outbreak of smallpox would threaten the Continental army, George Washington used the method to vaccinate his troops. Historians believe this may have had an important influence on the outcome of the Revolutionary War. Two decades later, Edward Jenner (1749–1823) developed the first true vaccine when he infected people with cowpox. It caused a mild disease but protected them from smallpox, which was much more deadly. The reason that it worked lay in the fact that the viruses that cause the disease are closely related and the immune system does not distinguish between them.

Pasteur began artificially weakening disease organisms in the laboratory for use as vaccines for the prevention of cholera, anthrax, and rabies. In 1885, he made a very risky decision to treat a young boy named Joseph Meister, who had been bitten by a rabid dog. The only treatment available was an experimental vaccine developed by a colleague, and it had only been tested in a few dogs. The boy's cure quickly elevated Pasteur to the status of a national hero. It also convinced the medical world that microorganisms caused illnesses and could sometimes be used to cure them. The germ theory of infectious diseases had come to stay, and it provided doctors with their first effective defense against epidemics that had long haunted mankind.

Still, they are not the solution to every infectious disease, at least not yet. Most modern vaccines are directed against viruses. The development of bacteria-killing antibiotics in the 20th century initially suggested that vaccines against bacteria might not be necessary. This was welcome because attempts to develop vaccines against tuberculosis and many other serious diseases had failed. Researchers have also been unable to develop treatments for HIV and many other viruses because, like some bacteria, they have mechanisms that help them evade the immune system. The AIDS virus slips into the white blood cells that should be fighting it and remains hidden there for a long time; bacteria and other parasites sometimes adopt disguises by changing the molecules on their surfaces. The vaccines needed to fight such clever parasites will have to be more sophisticated.

There is a renewed interest in making bacterial vaccines because many dangerous strains have developed resistance against common antibiotics. Finding cures will require a precise understanding of both the infectious agents and the immune system. The strategies that are being developed to do so may equally be useful in the fight against cancer and some types of genetic diseases. This theme is explored in more detail later in the chapter.

EVOLUTION, GENETICS, AND MATHEMATICS

In 1858, as Virchow was announcing his cell theory in Berlin, an even greater revolution was occurring in Great Britain. Two Englishmen were about to shake the foundations of what nearly everyone thought about the origins of life and its immense diversity around the world. Charles Darwin (1809–82) and Alfred Russel Wallace (1823–1913), living on opposite sides of the globe, had come to nearly identical conclusions. Just as each cell arose from a preexisting cell, each of the Earth's species had evolved from an earlier form of life, in an immensely long, unbroken chain stretching back to the first cell.

Evolution was based on a few logical principles that could be observed nearly everywhere:

- *Variation*—Each member of a species (except for identical twins) has slightly different features
- *Heredity*—Some of these unique characteristics are passed down from plant or animal parents to their offspring.
- *Natural selection*—In a given environment, some features give certain members of a species an advantage at survival and reproduction. If this bias continues over many generations, more and more of the population will be made up of their descendants, until they dominate the entire species.

Different environments—with unique climates, types of food, predators, and other factors—would favor different features. The genius of Darwin and Wallace was to see how, over long periods of time, tiny variations within a species could transform it into new ones.

There was little debate about the single points of the theory; what worried religious leaders and others was that, taken

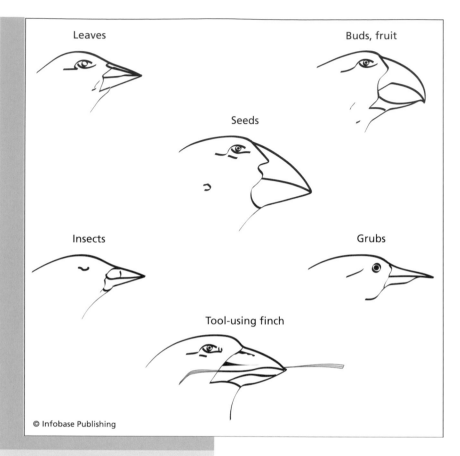

Leaves

Buds, fruit

Seeds

Insects

Grubs

Tool-using finch

© Infobase Publishing

Charles Darwin noticed small differences in several species of finches living on the Galápagos Islands in the Pacific—a perfect example of natural selection. Mutations caused small changes in the shapes of beaks, which allowed the birds to exploit different sources of food. Over time the birds whose beaks were best suited to their lifestyle survived better and had more offspring, and now specific beak-building genes dominate in each population.

together, they threatened the idea of a miraculous creation, in which humans and other species sprang fully formed from the mind of a creator. For the first time, scientists had a coherent theory to explain the diversity of life—and possibly even its origins—that could be tested in various ways.

Heredity was fundamental to the theory, but in the 1850s no one had any idea of how features were transmitted from parents to their offspring. Darwin's own speculations on the subject—that parents' traits were combined in a souplike mixture—were quickly proven wrong by his cousin Francis Galton (1822–1911). It was not a blow to the theory: Any mechanism for heredity would work, as long as it produced children who were neither perfect copies of their parents nor completely unlike them. If those conditions were met, selection would act on a species.

As the powerful framework of evolution began changing the way biologists looked at life, a reclusive monk named Gregor Johann Mendel (1822–84) was quietly going about solving some of the major questions about heredity. He succeeded at a problem that had stumped the greatest scientific minds of history for several reasons. First, he worked with peas and other plants in which reproduction could be carefully controlled. Second, he started from the assumption that various aspects of an organism—such as the shape and color of a seed—were separate features that were inherited independently of each other. Finally, he was talented at statistics, becoming one of the first researchers to apply mathematics to a biological problem. Exact numbers would be necessary to discover the laws that governed inheritance.

Mendel's work revealed that in most species, males and females each contributed one hereditary unit (later called a gene) to their offspring. These units might be the same or different—for example, both parents might pass along a gene that made peas yellow, or one might contribute a gene for yellowness while the other passed along greenness. Different forms of the same gene would be called *alleles.* For example, there was a gene that determined a pea's shape, with wrinkled and round alleles. A single plant might inherit two alleles for one gene. If that happened, Mendel concluded, one of them would be *dominant* and the other *recessive*. In peas, the green allele was dominant over yellow.

Darwin never read Mendel's work. The monk, on the other hand, was aware of evolution, but he did not see its immediate relevance to the questions he was investigating. Given the tools and concepts at hand, he could only look at one type of variation—alleles—and the focus of his work was why plant offspring were so similar to their parents, rather than where new traits might come from. To make a comparison: If the two men had been studying languages, Mendel would have been focused on small differences between the accents of two English speakers, whereas Darwin wanted to understand the relationship between French and Latin, or English and German. Different approaches were necessary, and this gap between genetics and evolution would not be closed for many decades.

Mendel did not live to see it happen. On the advice of another scientist, he tried to reproduce his results in another plant; it turned out to be difficult to handle and the experiments yielded confusing results. He began to doubt his own work just at the time he was appointed head of the abbey where he lived, and he spent the last years of his life immersed in its business.

The importance of his findings remained unappreciated until they were rediscovered at the beginning of the 20th century by three scientists working independently on similar problems in different countries: a Dutchman named Hugo de Vries (1848–1935), a German named Carl Correns (1864–1933), and an Austrian researcher, Erich Tschermak von Seysenegg (1871–1962). William Bateson (1861–1926), a British scientist, also played a vital role in bringing Mendel's ideas to the world and transforming them into a new science, which he called genetics. Bateson first encountered Mendel's name through a paper written by de Vries that he read while on a train to London. Bateson went on to demonstrate that Mendel's principles held true for animals as well as plants. He also believed they might provide a way of linking heredity to evolution.

This turned out to be more difficult than anyone expected, mainly because of the issue of variation. Genetics could explain part of it—for example, how two plants that produced green peas might give rise to a plant with yellow ones. But this happened because the trait for yellowness was already there. The focus of genetics was how existing alleles were shuffled from parents to their offspring. Evolution needed something more: an explanation for the appearance of completely new traits. Human beings were not just a peculiar arrangement of alleles that had once existed in bacteria and were simply being shuffled around in new ways.

De Vries proposed a solution with the concept of *mutations*: mistakes that occurred as genes were copied or during their passage from parent to child. Two decades later, American scientists would discover another type of change that could occur: Genes sometimes were duplicated, and offspring could inherit extra copies. This provided extra genetic material for evolution to work on and could partly explain how very complex organisms might arise from much simpler ones.

This discovery came from the research team of Thomas Hunt Morgan (1866–1945), who had established a laboratory at Columbia University in New York with the intent of catching evolution in the act. He planned to breed a species until mutations occurred, then study how the changes moved through the population. To do so, he needed an organism that reproduced quickly, had lots of offspring, and was easy to take care of. A colleague recommended the simple fruit fly *Drosophila melanogaster,* which could be raised in glass beakers and fed on mashed bananas.

For two years, Morgan and his students raised flies without discovering any mutations. He may have been about to give up when in 1909 they began to appear—subtle differences in the color of the insects' bodies and eyes. The fact that they were transmitted to their offspring along Mendelian patterns proved that genes were responsible. Each new mutation revealed the existence of a new gene—if a fly suddenly developed white eyes, it must have undergone a mutation in a gene that normally made them red. No one knew how many genes an animal had or how they worked, and Morgan's focus quickly shifted away from evolution toward these themes. Over the next three decades, his laboratory found dozens of new genes and identified their positions on chromosomes. It would be a long time before the cause of mutations would be understood—spelling changes in the chemical language of DNA—but the group discovered that genes could be duplicated, inverted, or undergo other types of changes.

Mutations seemed to be abrupt breaks in hereditary information, like computer files that had become corrupted. Bateson believed that a small number of such events could quickly give rise to new species, but initially many of his colleagues disagreed. Darwin had seen evolution as a very slow process that gradually bent, twisted, and stretched the existing features of organisms into new forms, rather than quickly replacing them with something different. For example, in a species of antelopes, some animals would inevitably have slightly longer horns than others. If longer horns made them more attractive to females or otherwise led them to have more offspring, then length would undergo positive selection over many generations. This vision was probably largely due to Darwin's fluid mixing hypothesis of heredity, which offered no explanation for the arrival of completely new features.

The debate arose from a misunderstanding about how genes functioned. Genes could encode quantitative features like an animal's height, weight, or the length of its horns; but a single mutation could also have very dramatic effects—giving a fish two heads or a goat an extra pair of legs. The reasons would not become clear until many years later, with the discovery of the connections between cell chemistry and embryonic development. In the early 20th century, the disagreement had to be resolved by mathematicians. Evolution had attracted their attention because it raised interesting statistical questions—for example, how a mutation that happened in a single organism could spread through a population. Many were skeptical that such single events could spread far enough to become visible to natural selection, especially since many mutations were recessive. This meant that two parents would have to have an allele before it appeared in their offspring. George Udny Yule (1871–1951), a Scottish statistician, predicted that dominant genes would multiply in a population and wipe out recessive traits. But the American geneticist William Castle (1867–1962) calculated that with no natural selection at all, the frequency of alleles in a population would remain stable over many generations.

Darwin had predicted that natural selection would work hardest when the pressure on a population was extreme—when a high number of predators lurked in the neighborhood, when populations exceeded the food supply, or when environmental conditions changed. In animals such as birds, the pressure would be high all the time—he calculated that in just 200 years, eight pairs of swifts could produce 10,000 billion billion billion descendants if nothing kept them under control. But what about animals that were less fertile? Darwin made some calculations based on one of the slowest-breeding creatures on Earth, the elephant. If a pair bred between the ages of 30 and 90 and had only six offspring, they would produce 19 million descendants after only 750 years. That was far beyond the real population, and elephants had been around for much longer. Natural selection was clearly working on them as well. The question was how to detect it.

The mathematicians Godfrey Hardy (1877–1947) and Wilhelm Weinberg (1862–1937) independently came up with the same answer, based on statistics and probability. Their formula to describe the spread of an allele, now called the Hardy-Weinberg rule, consists of the following steps:

1. Determine the frequency of specific alleles among the adults in a species.
2. Find out which types of adults mate with each other.
3. Estimate the frequency of alleles among their offspring using Mendel's ratios.
4. Discover how many of the offspring survive to reproduce.

The formula could be used to test hypotheses about populations and evolution. It verified William Castle's prediction that in a large population with random mating and no natural selection, the genetic makeup would stay the same over many generations. Recessive alleles could survive for a long time; they would not be wiped out by dominant ones. The method could also be used to detect natural selection. If the proportions of alleles in a population changed significantly over many generations, it was a sign that something was favoring some forms of a gene over others.

The Hardy-Weinberg rule still did not address the big question of whether evolution was driven by mutations or by very gradual changes in features. That problem would be tackled by

Ronald Fisher (1890–1962), a talented mathematician who developed an interest in evolution at an early age. As an undergraduate, he began thinking about ways to use mathematics to bring evolution and genetics together. In one project, he showed that evolution could not be primarily driven by new mutations that happened all the time; instead, once a change in a gene occurred, the new allele became part of a species' gene pool and behaved like any other allele. At the beginning it would be rare, but if it offered a reproductive advantage, it might spread quickly.

Calculating rates of species change required numbers, so Fisher invented a value called variance. In his 1930 book, *The Genetical Theory of Natural Selection,* he also introduced a concept of *fitness,* meaning a measurement of how well a species is adapted (or not) to its environment. For most species, which had been shaped by millions of years of natural selection, this value would be high. But it could change because species were molded by the environments of the past, rather than the present. Human fitness, for example, was the product of the hunter-gatherer lifestyle practiced for 99 percent of human history, rather than the circumstances of modern industrial society.

Fisher's contemporary Sewall Wright (1889–1988), an American geneticist, saw selection happening within a fitness landscape, an imaginary place of peaks and valleys. The purpose of this metaphor was to describe how selection could change a species' profile over time. Most individuals in a species would be near the peak—an allele that was best adapted to the environment, with other variants of a gene falling off in a slope, and some vary rare alleles in the valleys. But changes in the environment might suddenly favor different characteristics. This would shift the ideal position of the peak and cause the number of infrequent alleles to rise.

Another important figure in the coalescence of genetics and mathematics was John Burdon Sanderson Haldane (1892–1964), one of the most colorful figures of 20th-century science. During his youth, he carried out experiments with his father, such as studies of the effects of air pressure on the body, and he used himself to experiment on. This damaged some of his vertebrae

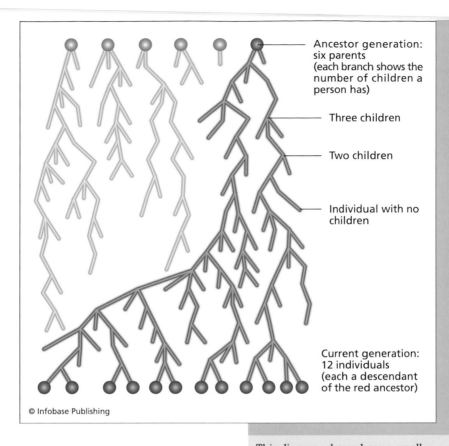

Ancestor generation: six parents (each branch shows the number of children a person has)

Three children

Two children

Individual with no children

Current generation: 12 individuals (each a descendant of the red ancestor)

© Infobase Publishing

This diagram shows how a small advantage in reproduction of some individuals and their descendants can have a big effect on a species over time. The family line at right (red) consistently has more children than the descendants of the five other ancestors at the top (blue). The differences may not be obvious in a single generation, but over time this family's genes may come to dominate an entire species.

and left him with a perforated eardrum. The latter was not a particular problem, he wrote: "The drum generally heals up; and if a hole remains in it, although one is somewhat deaf, one can blow tobacco smoke out of the ear . . . which is a social accomplishment."

In 1924, Haldane began a series of scientific papers called "A Mathematical Theory of Natural and Artificial Selection." He calculated that even if a trait offered only a very small reproductive advantage to the organism that inherited it, as little as 0.1 of 1 percent, with enough time it could become

The Failed Marriage between Evolution and Sociology

Although 150 years of research into evolution has con-firmed evolutionary theory in a lot of ways—including some that Darwin never envisioned—it is still widely misunderstood. The misconceptions are most obvious in the way people have tried to apply Darwin's ideas to human society. Evolution arrived on the scene at a time when industry and technology were rapidly chang-ing the Western world. Many people were obsessed with progress and believed that it would lead to a utopian society. Evolution brought the unwelcome messages that humans were not the ultimate goal of creation and that history was not driving their species toward physi-cal and moral perfection. It did not necessarily produce creatures that were more complex or intelligent than their ancestors, and species could fail—by becoming extinct.

Whereas evolution has been a unifying concept in sci-ence, helping to draw biology close to physics, chemistry, mathematics, and all the other disciplines described in this chapter, it has had a much harder time finding com-mon ground with the social sciences. Part of the reason lies with what was happening in the world of research and British society in 1858, the year that *On the Origin of Spe-cies* was published.

At that time the word evolution existed, but had a dif-ferent meaning, referring to the development of embryos into newborns and adults. That process clearly had a des-tination. A chicken egg produced a chick, and a baby in a human mother's womb gave rise to a human being. Her-bert Spencer (1820–1903), a British philosopher and po-litical and sociological theorist who was quickly becoming one of the most important thinkers of the 19th century, felt that human society was undergoing a similar developmental transformation.

Spencer described scientific progress as a method of trying out ideas and discarding those that did not work in favor of better ones. He proposed that the entire universe might behave the same way, moving from a state of simplicity to more complexity, from imperfection to higher order—not in a religious sense, but simply because natural laws worked that way. Since culture was a product of human beings, the laws that governed their biology must also dictate the development of culture. In an 1857 article called "Progress: Its Law and Cause," he wrote, "Whether it be in the development of the Earth, in the development in Life upon its surface, in the development of Society, of Government, of Manufactures, of Commerce, of Language, Literature, Science, Art, this same evolution of the simple into the complex, through a process of continuous differentiation, holds throughout."

When Darwin's book appeared, Spencer quickly became one of its strongest supporters. He believed it provided a firm biological foundation for his own ideas. But like many others, he was unable to give up the idea that humans were the pinnacle of evolution. He coined the phrase *survival of the fittest* to describe natural selection. Darwin was uncomfortable about this because he knew that fitness—the way Spencer meant it—was a loaded word. It assigned human values to the natural world and was a disguised attempt to marry concepts of improvement and progress to natural selection. In society, success meant the acquisition of wealth, power, and prestige. In evolution, success meant something completely different—living long enough to have more fertile offspring than other plants or animals of the same species.

(continues)

(continued)

Many people who read Spencer's books saw the parallels to Darwin's account of biological evolution without understanding the important differences, and this would have an important impact on the relationship between biology and sociology over the next century. Like most other philosophers, Spencer could not bear the idea that humanity's future was a matter of chance. Left on its own, without governments to intervene, human society would progress by favoring stronger and healthier individuals—not necessarily the rich, because he realized that poorer social classes were not really responsible for the conditions in which they lived. On the other hand, some of these people were clearly unfit through idleness or incompetence. Spencer believed that for such people to starve or suffer was a natural process.

It is not hard to see how these ideas could be turned against the poor, the sick, or groups that were considered somehow unfit by those in power. Spencer opposed charities and donations to the poor on the grounds that they ran against the principles of selection and promoted the survival of the unworthy. "The quality of a society is physically lowered by the artificial preservation of its feeblest member," he wrote. Helping defective people survive could harm society and perhaps even the human race. Taken to its extreme, this idea eventually led to *eugenics* programs—initiatives to control human mating. These programs were based on a complete misunderstanding of human genetics, and their disastrous consequences are discussed in the next chapter.

frequent in a population. He plotted what might happen after thousands of generations. A comparison of mathematical predictions to real measurements of allele frequencies would reveal how much of an advantage any particular mutation provided.

Haldane extended his calculations to unusual hypothetical cases, such as the following. Suppose that two recessive genes that affected eyesight were circulating in a population. On its own, each led to poor eyesight, but if someone inherited both genes, there would be an improvement. Haldane could predict whether the trait would survive and how frequent it might become.

By the 1930s, theorists had brought genetics and evolution together in what became known as the modern synthesis. But huge questions remained. No one knew what genes were made of, or why they led organisms to develop red eyes, five fingers, or any of their other features. In the meantime, geneticists and evolutionary researchers had begun to think about the implications of this new type of science on their own species.

CELL BIOLOGY, CHEMISTRY, AND GENES

If one had to pick a single icon to represent modern biology, it would surely be the DNA double helix: the spiral staircase–like ladder of sugars and nucleotides that contains the information needed to make a bacterium, plant, fly, or human being. This model of the molecule, proposed in 1953 by James Watson (1928–   ) and Francis Crick (1916–2004), revolutionized biology by showing how DNA could be copied and how mutations could arise—essentially proving, overnight, that genes were made of DNA. It also hinted at a way that genes might influence the structure and behavior of cells and, thereby, the formation of animal bodies.

The model could only be built because of a coming-together of chemistry, physics, and biology that had been under way for nearly a century. Until just a few years before Watson and Crick's discovery, many believed that proteins carried species' hereditary information, in spite of evidence in favor of DNA that had been accumulating since the late 19th century.

In 1866, the German biologist Ernst Haeckel (1834–1919) suggested that hereditary material might be found in the cell nucleus, whose function was unknown up to that time. Two years later, while trying to find a way to remove and study the nucleus, a Swiss biologist named Friedrich Miescher (1844–95) achieved the first extraction of DNA from white blood cells. Its chemistry was odd compared to that of other cellular molecules; for example, it contained large amounts of phosphorus, which was virtually unknown in other organic molecules. When Miescher submitted his results to a scientific journal, they were so unusual that the editor insisted that the experiments be repeated before agreeing to publish the discovery.

Microscopists were also gathering evidence that the cell nucleus played a key role in heredity. Oscar Hertwig (1849–1922), professor of zoology at the University of Berlin, looked at the huge, pearly white eggs of sea urchins and discovered that a sperm cell brings a new nucleus into the egg, which then fuses with the egg's own nucleus. The rest of the sperm is discarded. Three years later, his countryman Walther Flemming (1843–1905) stained nuclei with dyes and saw chromosomes for the first time. He watched as cells divided and discovered that the chromosomes were split up among the two new daughters. But Mendel's work was still unknown and without his data—showing that both parents contributed equally to heredity—the importance of these new findings was not immediately clear. Wilhelm Roux (1850–1924) and August Weismann (1834–1914), also German professors, figured out that fertilization is a process of combining chromosomes from each parent. These threads, Roux wrote, must contain the hereditary material, and he proposed that the information they contained was in a linear form, like the words of a text.

(opposite page) An organism's germ cells—sperm or egg—are among the first cells to develop in the embryo. They arise before the development of sex organs and migrate as these organs form. These images show the location of germ cells (yellow spheres) at various stages in the development of a fruit fly larva. Red shows the tissue that will become the fly gut, and green is tissue that will become sex organs. By maturity, the cells have taken up a position next to the future sex organs.

Weismann tried to pull all of these observations into a single theory. He believed that organisms maintained reproductive *germ cells* separate from the rest of their cells (which he called the soma), and this helped explained why organisms did not pass

along traits acquired during their lifetimes to their offspring. This idea was central to evolution but was still controversial among scientists, many of whom felt that natural selection was a severe and amoral system. Weismann put it to the test with an experiment in which he cut off the tails of mice for several generations in a row. If Darwin was wrong, he reasoned, the mice would eventually produce offspring with no tails. But this never happened. Neither behavior nor lifetime events affected the protected germ cells.

Weismann believed the material in these cells, which he called the germ plasm, would be the key to understanding heredity. Whatever the substance was, it was passed along intact from generation to generation, separate from the rest of the body. The soma was like a flower which grew and died within a year; the germ plasm was like the body of the plant, which survived season after season. The function of sex was to mix up the germ plasm of separate organisms, ensuring variety within species.

With the rediscovery of Mendel's work, the search for the physical location and the chemical nature of genes was ready to begin in earnest. At the turn of the century, the American geneticist Walter Sutton (1877–1916) proposed that genes were located on chromosomes. Working at the same time in Munich, Theodor Boveri (1862–1915) discovered that if more than one sperm managed to fertilize an egg, the resulting embryo had too many chromosomes, failed to develop, and died at a very early stage. This led him to an important conclusion: An organism needed not only a complete set of chromosomes, but also the right number. Through a series of experiments conducted between 1901 and 1905, he became convinced that each chromosome possessed unique qualities. Each contained a unique subset of the instructions needed to build an organism. Too many chromosomes meant too many instructions, and too few meant that important information was missing. The next task was to try to discover which information was stored where. An important first step came in 1905, when the American biologists Nettie Stevens (1861–1912) and Edmund Wilson (1856–1939) discovered that the X-Y chromosome pair contained the genes that determined an organism's sex.

If chromosomes contained only DNA, scientists would surely have realized much more quickly that genes were made of this molecule. However, the DNA in a cell nucleus is linked to a huge number of proteins in a mixture called chromatin. Many researchers were convinced that the genetic code was contained in the proteins, which were complex molecules built of 20 amino acids. The chemical language of DNA was much simpler, made up of only four nucleotides—too simple, perhaps, to produce complex organisms.

A major step forward came through the work of Frederick Griffith (1879–1941), a medical officer at the Ministry of Health in London. He was studying two strains of bacteria that were very similar, trying to figure out why one caused severe pneumonia in humans and the other did not. There was only one obvious difference: the infectious, smooth (S) form of the bacterium built a capsule around itself, while the rough form (R) did not. Griffith inoculated mice with a mixture of dead S-type and live R-type bacteria. He expected that the mice would stay healthy and the bacteria would die, because he had not injected the animals with any live infectious cells. But when he drew blood he found S-type bacteria that were alive.

Either the S type had somehow been brought back to life or something had changed the R bacteria into the S type. If the latter was the case, it meant that R bacteria were acquiring new hereditary information. Griffith began a new round of experiments to try to find out what this transforming substance was made of. One possibility was that fragments of proteins from the S bacteria were somehow being absorbed into R bacteria and were being used to build capsules, but Griffith had another idea. Rather than receiving building materials, the bacteria might be receiving the instructions to make the capsules. In other words, R bacteria had developed the capacity to make a new protein.

Griffith's investigations ended with his tragic death when London was bombed by the Nazis in 1941. But his work had attracted the interest of another scientist. Oswald Avery (1877–1955), a physician and researcher at the Rockefeller Institute in New York, was trying to develop a vaccine for pneumonia.

That work became unnecessary through the discovery of antibiotics, which very effectively killed the pneumonia bacteria. But the project gave Avery what he needed to follow up on Griffith's experiments, which looked like the most promising way to find bacteria's hereditary material. Members of his lab purified molecules from the S type and showed that DNA alone was able to transform the R type into infectious pneumonia bacteria. Avery cautiously proposed that in bacteria, DNA was the hereditary material, and that perhaps this was true of other forms of life as well. Yet other researchers remained skeptical.

One person who believed him was Erwin Chargaff (1905–2002), an Austrian working nearby at Columbia University in New York. He wrote, "Avery gave us the first text of a new language, or rather he showed us where to look for it. I resolved to search for this text." If DNA was truly the language of heredity, it could not be the same in every species, so Chargaff began trying to find differences in DNA.

He started out by simply comparing how much of each of the four bases could be found in yeast cells and the tuberculosis bacterium. By chance, he had chosen two organisms with major differences in composition of their DNA. Yeast had high amounts of A and T but much lower amounts of G and C, exactly the opposite of the bacterium. Chargaff tried the same thing with other organisms and found that each had its own particular recipe of DNA. In humans, for example, about 30.5 percent of DNA was A, 31.8 percent was T, 17.2 percent was C, and 18.4 percent was G. The tuberculosis bacterium gave a much different picture: 15 percent A, 13.6 percent T, 34 percent C, and 37.4 percent G.

The fact that each organism had its own recipe of bases meant that DNA might be the molecule of heredity. Chargaff noticed another curious fact: In any given organism, A and T were found in almost identical amounts; the same was true of G and C. Although he did not realize it, these numbers provided one of the most important clues as to how the DNA molecule was put together. It would not be explained until James Watson and Francis Crick understood DNA's structure.

PHYSICS, CHEMISTRY, AND GENETICS

Alongside Virchow's presentation of the cell theory and the first announcement of evolution, the year 1858 saw an important breakthrough in chemistry. Archibald Scott Couper (1831–92) and Auguste Kekulé (1829–96) drew the first blueprints of molecules: diagrams showing the positions of atoms and their relationships to each other. DNA, RNA, and proteins are fundamental units of life, but the atoms that make them up are even more basic. As any engineer knows, the function of a machine depends on the way its parts are assembled, and the same is true for molecules. By the mid-20th century it had become clear that understanding genes would require learning about their chemical makeup and physical structure.

Chemists knew that DNA consisted of a sugar called deoxyribose, plenty of phosphate atoms, and the four nucleotide bases. Each component has a particular shape and chemistry that determine how it snaps onto the others. With very simple molecules, it is sometimes possible to guess how the parts fit together just by looking at the chemistry of the subunits, but in this case there were too many ways that the pieces might fit.

The details of DNA and other molecules such as proteins were too small to be seen through even the most powerful electron microscopes, so chemists were trying to understand DNA's structure by watching how other molecules changed it—a bit like ramming cars into each other to study their engines. *Crystallography* took another approach, turning molecules into crystals and exposing them to X-rays. This method, developed by physicists, had provided some important information about the shapes of proteins; perhaps the same thing would work with DNA.

When an X-ray beam passes through an object, some of the waves collide with atoms' electrons and are diffracted (they scatter off in a new direction). William Astbury (1898–1961) shined X-rays through molecules and captured the scattering patterns on photographic plates. Usually the resulting image was an unreadable smear. But a molecule whose atoms were arranged in precise, repeated groups scattered the waves in the same directions over and over again, creating a symmetrical pattern that hinted at the shapes of molecules. Astbury had been trying this with proteins that had formed crystals. In crystals, molecules are often arranged in precise lattices that repeat over and over again, billions or trillions of times. This creates the regular structures necessary to obtain a clear diffraction pattern.

Very pure DNA could either be made into crystals or pulled into fibers that also provided regular diffraction patterns. When Astbury examined DNA fibers with X-rays, he obtained some basic information about the size and architecture of the molecule. His interpretation was that the bases fit together into flat disks, squeezed very tightly together like dinner plates stacked in a column. He could measure the diameter of the disks and the height of each plate. However, many of the details remained blurred; without knowing it, he was working with two different forms of DNA. In his images they were superimposed.

The problem interested the great American chemist Linus Pauling (1901–94), who carried out similar experiments in his laboratory at the California Institute of Technology. He proposed a structure for DNA showing the molecule as a braid of three strands organized in a helix, like a spiral staircase with three handrails. It was one of the few times Pauling was wrong. Considered to be one of the greatest chemists of the 20th century, he had already used crystallography to study the composition of proteins; this work earned him a Nobel Prize in chemistry in 1954. Eight years later he became only the second person in history to win a second prize in a different category (the other was Marie Curie). This time it was the 1962 Nobel Peace Prize, for his efforts to stop the testing of aboveground nuclear weapons.

He paid a price for his political activities. In 1952, he had been denounced as a communist before the House Committee on Un-American Activities. When Pauling wanted to attend a scientific meeting of the Royal Society in London, he was refused a visa. One of his colleagues, Robert Corey, went instead. During the trip, Corey met with a young researcher named Rosalind Franklin (1920–58), who was also using X-rays to investigate DNA. It is hard to tell what might have happened had Pauling attended the meeting; he might have obtained data that would have led him to an accurate model of DNA. Franklin's work was about to play a crucial role in figuring out the molecule's structure.

Another incorrect model had just been proposed by the British scientist Francis Crick and his young American partner, James Watson, working in Cambridge, England. Watson had obtained his Ph.D. at the age of 22, working on viruses that infected bacteria at the University of Indiana, and had come to Cambridge determined to solve the riddle of DNA's structure. He was now 23 and Crick was 35, but the two men quickly recognized each other as two of the brightest people on campus and hit it off. They had a lot of catching up to do when it came to DNA; neither was an expert in chemistry. Their first diagram of the molecule was so wrong that it embarrassed their boss, Sir Lawrence Bragg, and he ordered them to stop working on it.

Meanwhile Franklin, an hour away by train in the laboratory of Maurice Wilkins in London, had solved a major problem regarding the X-ray images of DNA. She had figured out that DNA came in two forms: a dry and a wet form. Under humid conditions, more hydrogen atoms were packed into the molecule and that changed its shape. The preparations of the molecule made by Astbury and Pauling held both forms and caused blurring. Using only the B form, Franklin now obtained the sharpest-ever images of DNA and began trying to interpret what they meant about its structure, but interrupted the work to go on vacation. While she was gone, Wilkins showed some of her X-ray images to Watson. One look convinced him that DNA formed a double helix.

The problem that now faced Watson and Crick was like one of those wooden puzzles in which oddly shaped pieces have to be fit together to form a tight, geometric shape. In this case, the shape that had to be built was a helix, and the pieces were sugars, phosphates, and bases. Watson made cardboard cutouts in the shape of the four bases and began working on the puzzle. No matter how he tried to attach them to each other, something always bulged outside the helix. He was stuck until his office mate—ironically a former student of Pauling's—told him that bases existed in two different chemical forms, with slightly different shapes. Watson had been using a form with an extra oxygen atom, so now he remade the shapes without the oxygen. He was idly fitting them together when he had a sudden revelation: When A snapped onto T, it had almost exactly the same size as G fit to C. Fit together, their size matched the dimensions of the helix in Rosalind Franklin's X-ray photographs.

Watson showed Crick what he had discovered. They immediately realized that the steps of the DNA spiral staircase were the bases, rather than the sugars. Each step held either an A combined with T, or a G with a C. The steps were connected by winding rails of deoxyribose sugars (the backbone). Between each of the steps, there was a slight twist, making the whole structure into a helix rather than a straight, ladder-like column. They quickly wrote a paper called "A Structure for Deoxyribose Nucleic Acid" and submitted it to the journal *Nature*. It was published three weeks later—an amazingly short time, given the fact that it first had to be read and commented on by experts.

This brief article would revolutionize biology because the molecule's building plan provided immediate insights into its behavior. It explained Chargaff's discovery that A and T occur in identical amounts in an organism, as do G and C. The pairing of the bases revealed how DNA might copy itself. If the two strands of DNA were split apart, each base would attract and link up to just the right partner nucleotide, creating a second strand. The article even suggested a way that mutations could occur, in spite of the fact that bases formed regular pairs. In rare cases, hydrogen atoms might bind differently to a base, slightly changing its shape. As one strand was copied, it might then attach to the wrong base.

A crucial point was that any sequence—any possible spelling of the four bases—formed the same shape. A long strand made up only of As joined to Ts would create the same double helix as a sequence consisting only of G-C pairs. Each organism could have its own DNA sequence; a language with four letters was rich enough to create all the diversity of life on Earth. Evolution was built on a single scaffold.

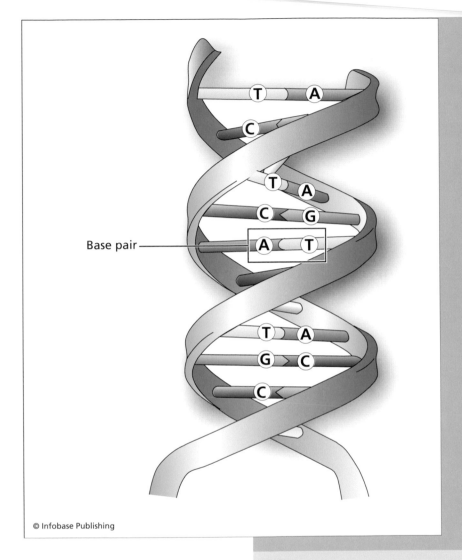

Base pair

© Infobase Publishing

With this single, power-ful image, some of the most important questions about genes, cell replication, and evolution were resolved, all at once. Nine years later, Watson, Crick, and Wilkins were awarded the Nobel Prize in

Watson and Crick's double helix model of DNA, based on data from physics and chemistry experiments, showed that bases bind in complementary pairs. This demonstrated that genes were made of DNA, explained how the molecule could be copied, and also suggested a reason for mutations.

An image of the Brookhaven National Laboratory campus in Upton, New York, as it will appear after the construction of the National Synchrotron Light Source II (NSLS II), the circular structure in the background. In the foreground is the current NSLS. *(U.S. Department of Energy)*

medicine or physiology for their discoveries. Franklin had died of cancer in 1958 and was therefore ineligible.

X-rays have continued to play a vital role in molecular biology ever since. Enormous *synchrotrons* (particle accelerators), built for physics experiments, have been harnessed to provide high-energy X-rays. Most of the projects aim to obtain high-resolution structures of proteins in crystal form. Knowledge of protein shapes has become essential in discovering their functions in health, disease, and the activity of drugs.

GENETICS, EMBRYOLOGY, AND EVOLUTION

An organism does not inherit features (such as blue eyes) fully formed from its parents. Instead, it inherits a genome that tells a single cell (the fertilized egg) how to specialize and build tissues and organs. During the nine months between fertilization and birth, they arrange themselves into tissues and complex organs such as the eye. The goal of embryology (now known as developmental biology) has been to understand the processes by which genetic information is transformed into a body.

Researchers have tried to accomplish this in two main ways: working from the developed body back down to the level of cells and molecules, and working from genes upward by studying their functions in cells and tissues. Only in the last few decades have these approaches truly found common ground with the identification of genes that help shape the embryo's body as it grows.

The earliest embryologists were physicians whose main method was to dissect fetuses that had miscarried at various stages of development. The similarity between the bodies of humans and other animals meant that a great deal could be learned from dissections of animal embryos as well. Already at the beginning of the 19th century, comparative anatomy was used to study adult animals, revealing surprising similarities between body parts such as the bone structure of human arms and hands, the wings of bats, and the flippers of whales. Karl Ernst von Baer (1792–1876) extended this work to embryos and made the discovery that animals that looked quite different as adults often went through embryonic phases in which they looked remarkably alike.

Evolution offered a possible explanation—organisms had inherited similar features (called *homologues*) from their common ancestors. This is still an important concept in evolutionary theory. Homologues appear at every level of biological organization. Related genes produce similar body structures in a huge range of species.

One of Darwin's most enthusiastic followers, the German researcher Ernst Haeckel, became known for an interesting attempt to unify embryology and evolution. Haeckel was born in Potsdam, near Berlin, and received a degree in medicine before deciding that he was cut out more for a life of research than one of dealing with sick patients. Haeckel followed in the footsteps of von Baer, armed with better microscopes and the new theory as a framework for his observations.

As he compared embryos of many species, he developed a radical new hypothesis called *recapitulation*. He believed that as an individual organism undergoes development (*ontogeny*), it retraces the evolutionary history of its species (*phylogeny*). All life began as a single cell; so does an individual. The earliest multicellular life-forms were probably ball-shaped, with just a few different types of cells; a human embryo goes through a similar phase. Only in later stages of development do animal embryos start to look markedly different from each other. For Haeckel this reflected the fact that most of today's species arose recently in evolutionary history.

Haeckel found fascinating evidence for his claims. At one stage, a human embryo develops structures like gill slits that then disappear again. This only made sense, he said, in light of the fact that the distant ancestors of mammals were fish. Haeckel drew images of embryos at various phases to show how similar their body plans were. This work has been criticized because his drawings tended to emphasize the similarities rather than the differences among embryos. Haeckel's defenders point out that such a critique is easy to make in the days of photography, where objective images can be made of samples. Drawing is always subjective; an artist must make choices about which features to emphasize after looking at many specimens, and there is always a danger of wishful thinking creeping into the process.

The recapitulation hypothesis was in many ways logical and appealing. Knowing nothing of genes or DNA or their roles in shaping organisms, researchers were struggling to understand how one species might be transformed into another. It was easy to imagine that this could happen when a species added on developmental stages or its development slowed down. But Haeckel took the idea much further, claiming that "ontogeny recapitulates phylogeny" was a law. A human embryo did not simply resemble that of a fish; he believed that it actually became a fish—an adult fish—on its way to becoming an adult human. The hypothesis claimed that evolution worked by adding new developmental stages to the end of an animal's life.

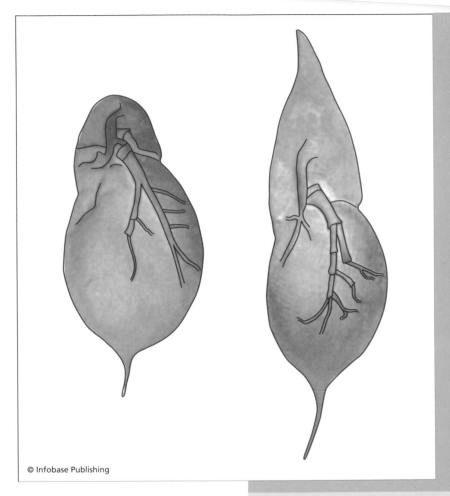

This would soon be revealed as a serious flaw in the hypothesis. Natural selection works at every stage of an animal's life to shape it, which means that the embryonic phases of an organism's growth also can be shaped in ways that are unique to a species. Fritz Müller (1821–97), a German expatriate living in Brazil, studied crustaceans to prove that evolution shaped larvae as well as adults. As free-swimming organisms, the larvae of each

Evolutionary adaptations can take place at any stage of development. A water flea larva (left) develops differently when predators are nearby, growing a larger, helmetlike head (right) that makes it harder to swallow. This probably happens because it senses the predator's molecules in the water.

species would have to cope with predators and other features of the environment, so natural selection would shape them just as it affected adults.

In the meantime, numerous examples of such embryonic or larval adaptations have been found. Species such as water fleas, frogs, and carp develop differently when predators are nearby. Water flea larvae grow larger, helmet-shaped heads that make them harder to swallow. Tadpoles grow stronger tails that allow them to swim more quickly and make faster turns.

Despite its flaws, the recapitulation hypothesis encouraged scientists to begin thinking of development in evolutionary terms and to focus on the processes by which hereditary information directed the growth of body structures, rather than only end results such as fully formed limbs. If the bones in a dolphin fin could be matched one-to-one with the hand of a primate, the processes that created the bones should also be homologous. This could be followed all the way back to the earliest stage of embryonic differentiation: *gastrulation,* in which embryos of nearly all animal species develop three specialized layers. But without an understanding of the genetic code and its relationship to the molecules in cells, scientists were stuck there.

A few decades later, Walter Garstang (1868–1949) and Gavin de Beer (1899–1972) pointed out the importance of timing when comparing embryos of different species. New species did not usually arise by adding on developmental stages, as Haeckel had proposed. Instead, each organ and bodily system should be looked at as an independent module. The development of one part might speed up compared to the others, a bit like the way engineers make changes in computers. They might develop a new graphics or sound card while the rest of the machine stays the same. Of course, this may well put pressure on other parts of the machine to change—new games might be made to take advantage of the features of the graphics card and for the games to run well, they might require more RAM or changes in the keyboard. The fact that one change often prompts others could explain why a new species had longer limbs than its ancestors or why humans and chimps do not have tails.

Recapitulation is undergoing a sort of limited revival in the molecular view of evolution. Today it might be phrased in this way: "Organisms resemble each other at many stages of development because the genes they inherited from common ancestors work in a similar way to create the same kinds of body structures." This happens even when the starting points and ending points of development are different—the eggs of a chicken and human are quite different, and they are very different as adults, but particular phases of embryonic development are similar.

A few species have indeed evolved the way Haeckel believed, a process that evolutionary researcher Stephen Jay Gould (1941–2002) called terminal addition. In this process, a new species adds developmental stages beyond those of its ancestors (like adding boxcars to a train). In other cases, evolution has sent species like caterpillars off on a path that is completely different from other kinds of larvae, like trains leaving a station in different directions. And a few types of organisms underwent the opposite of what Haeckel proposed, becoming stuck at an early phase of development because of changes in the genes that were supposed to trigger the next step. An example is the axolotl, a rare salamander found only in Lake Xochimilco in Mexico. This creature remains in its larval form its whole life long and can even reproduce without becoming an adult.

A bottom-up approach to development has only become possible recently, thanks largely to the work of the German researcher Christiane Nüsslein-Volhard (1942–   ) and the Americans Eric Wieschaus (1947–   ) and Edward Lewis (1918–2004). Their work on the fruit fly in the late 1970s and 1980s created a new kind of developmental biology that was strongly tied to molecular biology and led to their sharing the 1995 Nobel Prize in physiology or medicine. This was so important because embryology had not yet truly come into the molecular age.

Thomas Hunt Morgan's lab, where classical genetics was born, had focused almost entirely on the appearance of traits in adult flies, hoping to identify the genes that were responsible. It was a good approach at the time, given that scientists knew almost nothing of what genes were made of and how they

While investigating genes that help establish body structures in fly embryos, the German researcher Christiane Nüsslein-Volhard and her colleagues changed the relationship between developmental biology and genetics. In the intervening years, the zebrafish has become one of biology's most important model organisms due to Nüsslein-Volhard's efforts. *(Association of German Foundations)*

functioned. Morgan himself was not particularly interested in the chemical nature of genes; at the time, biochemistry was not far along enough to answer the important questions. The discovery of DNA's structure and the birth of molecular biology completely changed this situation: The biochemistry of the cell was now the central theme of biology. Work with flies was regarded as old-fashioned.

This explains the skepticism that greeted Nüsslein-Volhard and Wieschaus when they arrived at the European Molecular Biology Laboratory in Germany in 1979. Their research plan was to try to identify genes from the mother fly that influenced the development of embryos. To create mutations they fed male flies sugar water containing substances that damaged DNA, then allowed them to mate with females. This often produced malformed embryos, a starting point for discovering which genes had which effects. The work required Nüsslein-Volhard and

Wieschaus to spend several months peering at embryos under a microscope with two sets of eyepieces, looking for developmental defects.

The project paid off quickly. The fly embryo turned out to be ideal for the new approach: Its body is divided into stripelike segments that disappear or become rearranged through mutations. Since each segment gives rise to particular body structures that develop later, the mutations served as a guidebook to genes crucial to the development of flies and most other animals. These genes, some of which are now known as *HOX genes,* play an important role in establishing both the overall building plan for a body and the structure of specific parts, like arms and legs. They are ancient and so important that they have been preserved throughout the animal kingdom.

THE RISE OF GENETIC ENGINEERING

As Nüsslein-Volhard, Wieschaus, Lewis, and a handful of other geneticists were planting the seeds of a new type of developmental biology, a much louder revolution was beginning in California. Barely a century after Mendel's discovery of genes, scientists had learned to read the genetic code. This accomplishment was also a triumph of blending chemistry with biology. During the 1970s, Frederick Sanger (1918–), a British biochemist, developed a method to sequence DNA. The accomplishment earned Sanger the 1980 Nobel Prize in chemistry. It was the second time he had won the prize. The first time, in 1958, came for a method of determining the amino acid sequences of proteins.

Reading the genetic code set the stage for learning to write in it through genetic engineering: a set of powerful new tools to study and manipulate organisms' genes. Genetic engineering allows scientists to alter the DNA of a cell, plant, or animal in deliberate ways for research purposes, so that they can observe how changes in genes affect an organism. This is called *reverse genetics* because it is the opposite of the classical method of starting with a phenotype and looking for the gene that is responsible (*forward genetics*). By the end of the 20th century, it had become routine to make targeted changes in plants, animals, and human cell lines. Genetic engineering also led to the development of applications such as new foods and the use of microbes and animals to produce molecules such as insulin, used in medicine.

DNA naturally undergoes mutations; starting in the 1920s, scientists began learning methods to alter the molecule artificially. Hermann Muller (1890–1967), a former assistant in Morgan's lab, showed that radiation could increase the rate at which mutations occurred. This was soon followed by the discovery of other *mutagens* (such as chemicals) that could accomplish the same thing. As useful as these techniques were, they all had a drawback: The changes they caused in genes were random and unpredictable. Sometimes it was impossible to connect a change to a specific gene. Researchers dreamed of a day when they could pick a gene, *knock it out,* and then study its effects over the course of an organism's development and lifetime.

A few key discoveries set the stage for genetic engineering, which would give them that opportunity. In the late 1950s, a Swiss scientist named Werner Arber (1929–   ) was investigating how bacteria become resistant to attacks by viruses called phages. Salvador Luria (1912–91), working at the Massachusetts Institute of Technology, had discovered that bacteria had proteins called restriction enzymes that helped protect them from the virus. Arber and Hamilton Smith (1931–   ), of Johns Hopkins University, showed that the proteins formed part of a bacterial defense system that chops foreign DNA into small pieces.

Genetic engineering requires a pair of molecular scissors (to remove a gene from one place) and a sort of glue (to paste it in somewhere else). Restriction enzymes provided the scissors. Bacteria contained another type of molecule, called a ligase, which could spot broken ends of DNA and mend the cuts. Organisms need such enzymes because DNA sometimes breaks by mistake. Ligases can scout the molecule and make repairs by matching the broken ends to matching sequences in a chromosome and gluing them into place. Sometimes they insert the fragment in the wrong place. If that happens to be in the middle of a gene, the information gets scrambled. So ligases provided a tool to interfere with existing genes. They can also be used to insert a new gene into an organism's DNA.

In 1972, Janet Mertz and Ron Davis of Stanford University combined restriction enzymes and ligases in a technique now known as DNA recombination. A year later, Herbert Boyer of University of California, San Francisco (UCSF) and two colleagues at Stanford University, Stanley Cohen and Annie Chang, put the method to work to move a gene from one species to another. They combined genetic material from a virus and a bacteria and inserted it into another bacteria. This artificial gene was taken up by the cell and used to create a foreign protein.

Modern techniques allow scientists to remove a gene, substitute another one for it, or add a new molecule to an organism's genome. These methods are now used routinely in medicine, agriculture, and all kinds of biological research. One application, for example, has been to use bacteria or other species of animals to make human insulin needed to treat diabetes. People with type 1 diabetes are unable to produce insulin, which is needed to regulate the body's uptake of glucose. Healthy people cannot serve as donors, because insulin cannot be obtained in large amounts from the body. Doctors used to administer insulin extracted from pigs or cows, but their bodies produce a slightly different form of the molecule that sometimes causes rejection by the immune system or long-term health problems. Changing the recipe of the animals' insulin genes through genetic engineering causes them to make a more human version of the molecule that can be safely used, without such side effects.

Genetic engineering has many other practical uses. Bacteria are being put to work for *bioremediation*—clearing the environment of contaminants such as pollution. In many cases, this happens naturally. Certain species of microorganisms are able to digest toxic substances such as oil. Researchers discovered that after an oil spill a thin layer of bacteria may form a biofilm, a sort of floating mat on the surface of the ocean, which breaks down the contaminants into less harmful substances. Scientists have been able to identify some of the organisms that are responsible and hope to learn to engineer the genes that permit them to do so.

Today, a rising percentage of corn, tomatoes, soybeans, rice, and dozens of other crops produced across the world have been manipulated through genetic engineering. As well as attempting to improve the size, taste, shelf life, or nutritional value of crops, scientists have transplanted genes that help protect plants from insects, fungi, and other parasites. These changes might help farmers ward off pests without the dangerous side effects of pesticides. On the other hand, growing numbers of ecologists in the environmental movement protest that genetic engineering might upset delicate balances in nature. These issues are discussed in more detail in the next chapter.

COMPUTERS AND BIOLOGY

The revolution in biology that started in the second half of the 20th century has depended on an equally amazing revolution in computers in a significant way. It is not surprising that the two fields have found a great deal of overlap. First, many of today's experiments produce huge amounts of data that cannot be captured or stored without computers. Analyzing it may involve comparing billions of pieces of information to billions of others—also impossible without the help of machines. Another use of computers is to solve extremely complex biological puzzles, such as looking at the amino acids that make up a protein and predicting the shape they will form, which is an important clue to the molecule's functions.

Now, during the first decade of the 21st century, researchers are using computers to model and simulate biological processes. Limitations in technology restricted early molecular biologists to observing a few components of a biological process at a time; today, the same events in cells are seen as taking place against a background of networks of molecules that shift from one state to the next in very complex ways. It takes a computer model to keep track of the components, let alone predict how the entire system will change when something happens. For all of these reasons, computers have become indispensable to biological research.

Previous sections of this chapter have introduced some of the early uses of mathematics and models in genetics. Gregor Mendel's talents at statistics were essential in allowing him to analyze inheritance in peas and other plants. The method exposed the dominant or recessive nature of alleles and other aspects of the behavior of genes. Thomas Hunt Morgan's laboratory used the rates at which genes were inherited together to map their locations on specific chromosomes. When the first computers were built, they were put to work on biological questions: Millions of calculations were needed, for example, to interpret the diffraction patterns produced by X-ray experiments, aiming to uncover the structure of protein crystals.

With the arrival of DNA sequencing, computers became essential partners in everyday biology. In 1982, Greg Hamm and Graham Cameron, two researchers at the European Molecular Biology Laboratory (EMBL) in Heidelberg, Germany, proposed collecting DNA sequence data in a universal public database. The project they created is now known as EMBL Bank and is hosted by the European Bioinformatics Institute (EBI), a bioinformatics center run by the EMBL near Cambridge, England. At nearly the same time, an American group created a database called GenBank, originally hosted at the Los Alamos National Laboratory, now moved to the National Center for Biotechnology Information in Bethesda, Maryland. The two groups and another project in Japan worked out a system for exchanging data to offer a free, worldwide, up-to-date service that researchers across the world can access online. As of February 2009, the databases contained 100 gigabases of information about DNA, RNA, and protein sequences.

These resources were in place in the late 1980s as high-throughput DNA sequencing became common, and the United States and other countries decided to embark on genome projects. Biocomputing methods were required to collect and assemble the information. Early in the Human Genome Project, for example, it was decided to take a shotgun approach to obtaining DNA sequences: Rather than starting at the beginning of the first chromosome and analyzing the DNA to the end in a linear way, the sequence was broken down into fragments of several hundred nucleotides. This sped up the project because it allowed parallel processing of many regions of the genome, but it required an additional step of assembling the fragments into a linear text. Each sequence had a bit at the end that overlapped the ends of sequences from other fragments; by matching up the ends, the computer could join the small bits of sequence into a whole. Accomplishing this required several days of constant processing on a farm of 100 Pentium III computers dedicated solely to the task.

Then came the next challenge: to search the sequence for genes. Protein-coding regions make up less than 2 percent of the human sequence. There are three main ways to identify them. If an RNA or protein has been found in an experiment on human cells and researchers have obtained its sequence, they can work backward from the protein's amino acids or the RNA's nucleotides and predict what the sequence of the gene must be. Then they simply scan the genome for the matching code. But some molecules are only produced for short periods of time in specific tissues and have never been detected in experiments. Here a second method comes into play: Years of work have revealed some standard features of genes that can be detected by computer programs. In bacteria, for example, most genes begin with one of a few types of promoter sequences that are usually similar to each other and relatively easy to find. Eukaryotes such as humans, animals, and plants have evolved more complex promoters and genes that have to be spliced—meaning that protein-encoding information is hidden between long stretches of nonsense called *introns*. This makes human genes much harder to find.

A third approach is based on evolution and uses sequence information from one species to detect genes in another. Experiments in flies, worms, mice, and dozens of other laboratory organisms have revealed tens of thousands of genes. Most of them have homologues in other organisms. Despite the fact that they have undergone mutations and other types of changes, a clever computer program can still detect the similarities. Thus, every new gene found in another species is compared to the human genetic code.

One of the most interesting uses of bioinformatics has been to answer questions about the history of life. Darwin's principle of common descent dictates that if two species share a feature, this usually means that their common ancestor also had the feature. The same is true at the level of genes: If two species have nearly identical DNA sequences, it is because they inherited the gene from a common ancestor. This method can be used to reconstruct the ancestral gene. Each living species may have a different version of it, but by comparing the spellings of enough versions, it is often possible to reconstruct the original sequence.

In 2005, Detlev Arendt's laboratory at EMBL used this method to compare the evolution of gene structure in animals. The scientists showed that the genome of the fruit fly is not only quite a bit smaller than the common ancestor of insects and vertebrates—its genes are also less complex. The average human gene contains 8.6 introns; in the fruit fly, the average is only 2.3 per gene. Originally scientists believed this meant that genes have been getting more complex over the course of evolution—humans are more complex than flies by most ways of measuring things and thus are often thought of as more evolved.

Arendt points out that humans are evolving at a much slower pace than insects, partly because of the speed at which they reproduce. Fruit flies reach sexual maturity two weeks after birth, compared to 14 or 15 years in humans (who usually wait longer than that to reproduce). Thus in just 250 years, flies have given birth to about the same number of generations as humans have produced in 100,000 years. Since every new generation is an opportunity for mutations and other changes in DNA to creep into the genome, flies are evolving faster.

Arendt's study compared the introns of humans, flies, and a worm that resembles their last common ancestor. His conclusion was that ancient genes were more like those of humans than flies—they had more introns. Here, evolution has not been making things more complex; it has been simplifying things by cutting out introns in fly genes.

Sequencing gave researchers a way to carry out a complete inventory of a cell's DNA sequences; it has been much more difficult to conduct a similar census of proteins. In the 1990s, scientists figured out a way to begin, using a method called mass spectrometry, adapted from chemistry. Proteins are extracted from a cell and are cut into small fragments by enzymes. These are shot past a magnet and fall into a detector, which essentially weighs them. The amino acids that make up the fragments carry different electrostatic charges, so the magnet bends them in different ways depending on their composition. The size and charge of a fragment determines where it falls. By analyzing this position, a computer program can figure out the amino acid recipe. The computer can compare this to the human genome sequence and identify the genes that match the protein sequence. The process is a bit like rearranging fragments of sentences into a text. It is dependent on the computer's ability to assemble the data in a meaningful way.

Computers also play a central role in the design of new drugs. When researchers have identified a target molecule—such as a protein or RNA that plays a central role in a disease process—they begin looking for a substance that can alter its activity, usually by docking onto it. This is normally done by screening a library of substances (known drugs and other small molecules) against the purified protein in the test tube, or cells that contain it. Sometimes computer "docking" programs are used to preselect likely candidates, through a puzzle-building-like effort that involves matching the surfaces of molecules on the screen. The real work for computers begins when a substance is found that alters the protein's activity in the desired way. It usually needs to be rebuilt to fit the target more precisely and to have stronger effects. Programs analyze its binding sites and recommend small structural changes that should make it more effective.

The most sophisticated use of bioinformatics methods is to model complex, dynamic systems in cells and organisms. In 2000, Eric Karsenti, a cell biologist at EMBL, began using models to try to understand the *microtubule* system in cells. Microtubules are fibers made of single protein subunits called tubulin. The

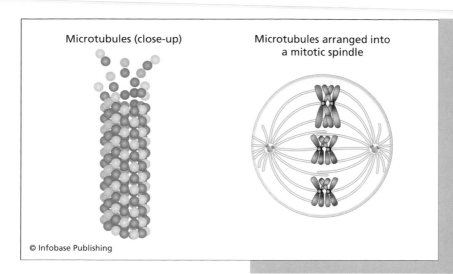

Microtubules (close-up)

Microtubules arranged into a mitotic spindle

© Infobase Publishing

The left image shows the structure of a single microtubule, built of two types of protein subunits called tubulins (red and green). For most of the cell's lifetime they form a sprawling network of fibers aiming outward from the border of the nucleus. During cell division (right), they form a spindle-shaped structure that pulls chromosomes in opposite directions to create two new daughters.

cell builds them by stacking tubulin in long columns—like stacking styrofoam cups— and then gluing several stacks together to form a tower. At the top, proteins are continually added and removed. Other molecules determine how quickly this happens and how high the stack becomes.

Normally, microtubules sprawl through the cell and are used as a sort of scaffold that provides structural support and gives the cell an overall shape. They also serve as traffic ways along which molecules are delivered through the cell. But during cell division, the entire system is completely broken down and rebuilt into a mitotic spindle. Microtubules form a double set of towing lines that stretch from poles on opposite sides of the cell to the center, a bit like two people standing on opposite sides of a field, holding thousands of strings attached to a fleet of kites flying in the air between them. The kites are chromosomes, and when the two fliers reel in their lines they separate DNA into two equal sets. Each set will become the nucleus of a new daughter cell.

Cells have neither a brain nor a master architect to direct this rebuilding project, and Karsenti wanted to understand how the components of the microtubule system were capable of self-organizing into such dramatically different structures. One discovery was that motor proteins play a crucial role. These are molecules that drive along the surface of microtubules, dragging along cargos on a flexible tether. Motors usually have two feet—regions that bind to the microtubule. Each time a foot lands, it undergoes a chemical transformation that causes it to let go again and then move along to the next foothold. Sometimes the feet bind to more than one microtubule, which pulls the fibers into alignment.

In collaboration with Stan Leibler at Rockefeller University, Karsenti and his colleagues have modeled this behavior in the computer to show that complex cellular structures such as the spindle can be generated by a few molecules, following a few simple rules generated by their physical structure. In making the model, postdoctoral fellow François Nédélec and the rest of the team had to know the following:

- the probability that a motor will bind to a microtubule or detach
- concentrations of the molecules that are involved
- the rate at which a microtubule grows and the amount of flexibility it has
- the probability that a motor will fall off the end of a microtubule
- and the speed at which a motor travels and its direction

Careful studies of cells gave the researchers the data they needed to establish some of these parameters, and their simulation of virtual mitotic spindles on the computer screen eerily mimics what they see under the microscope. The lab can now use the model as a way to test hypotheses about the behavior of microtubules and motors. Slightly changing one parameter on the screen—for example, telling a motor protein to move more slowly—may cause the spindle to break down and microtubules to assume a completely different shape. If the scientists notice a similar rearrangement in cells, they can look for a protein or something else that is slowing the motor down.

Such blends of computational and experimental science are increasingly being used by researchers in order to understand what happens in the cell and the body. A new term has been invented to describe this type of science—*systems biology*. At the moment, this phrase is used in so many contexts to describe so many types of work that it is hard to provide a simple definition. The reason it is being used so widely is that many feel molecular biology has entered a new phase, one in which mathematical models and computer programs are essential partners in the investigation of complex biological phenomena. This is a natural evolution of the trends described in this chapter. From their origins as separate fields in the mid-19th century, evolution, genetics, cell biology, chemistry, physics, medicine, and mathematics have become interwoven in a unified science of life. The next chapters show where this evolution is likely to lead and the ways that society is trying to cope.

2

Literature, Culture, and Social Perceptions of Science

Discoveries in biology since the 19th century have dramatically changed people's views of life, human nature, and the environment. Humans are now seen as both a product and part of the natural world. The body is viewed as an assembly of cells that grows in a systematic way from a fertilized egg, rather than as a mass of tissue formed all at once through a special act of a creator. The genetic code preserves a record of the entire history of evolution, revealing a link between people and every other living organism on Earth.

When these principles were first discovered, they ran so counter to what nearly everyone had previously believed that they provoked strong reactions in society. Religious institutions, which claimed to be the authority in matters regarding human nature, were concerned about the implications of a materialist view of life—and were afraid of losing their power. Other fears soon followed. Discoveries in the other sciences—especially physics—led to the development of terrible weapons. New types of technology such as television, personal computers, the Internet, and cell phones began changing people's daily lives, often so quickly that there was little time to adapt. Biologists and doctors predict that data and tools from molecular biology will have an equally profound effect on society.

Writers and artists have a unique opportunity to explore these themes by imagining future worlds or describing fictional situations in which science has changed society. Their visions have had a deep impact on how people think of science. This chapter demonstrates how, as scientists have constructed a new view of life, writers and thinkers have explored its implications.

FRANKENSTEIN: OR, THE MODERN PROMETHEUS

In 1816, Mary Godwin (1797–1851) traveled with her future husband, the poet Percy Bysshe Shelley (1792–1822), from England to Lake Geneva, in Switzerland. The couple was recovering from the death of a prematurely born daughter. They brought along Mary's stepsister, Claire, planning to spend the summer with George Gordon Byron, or Lord Byron (1788–1824)—famous for his poetry and infamous for a number of scandalous romantic affairs. The young group was brought together by their love of literature as well as their social ideas: All were concerned by the disturbing political and industrial changes taking place in their home country, England. It was not a good time for social radicals in Great Britain. The French Revolution, which had begun in a liberal spirit, had toppled its nation's monarchy through murder, pushed Napoleon to power, and embroiled Europe in war. England was in a conservative mood, and the group sought refuge on the continent.

The summer was gloomy and rainy, and they were often confined to the house for days at a time. They wrote, talked about science and politics, and shared ghost stories by candlelight. Byron suggested that each of them write a story about the supernatural. His own contribution was a vampire tale, based on stories he had heard during his travels through eastern Europe. Another guest, John Polidori, reworked the fragment into a story called "The Vampyre," which inspired a long romantic tradition of tales about the creatures.

Among the topics of discussion were the works of Erasmus Darwin (1731–1802), poet, physician, and natural scientist (grandfather of Charles, at the time a young boy). The elder Darwin had speculated on the possibility of bringing the dead back to life. The discussions stayed with Mary, and a few days later, still haunted by the death of her daughter, she had a vivid dream:

> I saw the pale student of unhallowed arts kneeling beside the thing he had put together. I saw the hideous phantasm of a man stretched out, and then, on the working of some powerful engine, show signs of life, and stir with an uneasy, half vital motion. Frightful must it be; for supremely frightful would be the effect of any human endeavour to mock the stupendous mechanism of the Creator of the world.

This dream formed the seed of the novel *Frankenstein* published in 1818, destined to become one of the most widely read books in the world and the inspiration for dozens of films and an entire genre of literature. Mary Shelley's book would establish one of the main themes of science fiction—the scientist driven by ambition or personal motives to create something without regard for ethical concerns. Then, whatever he creates escapes his control and wreaks havoc and terror on the public. In *Frankenstein,* the creation is a human being; in other novels it would be robots, microbes, or machines transformed into terrible weapons. Researchers are constantly seen as fall-

Mary Shelley, author of *Frankenstein*, in a portrait by Richard Rothwell *(National Portrait Gallery, London)*

ing prey to hubris, overweening pride, or reaching beyond one's grasp with no concern for the consequences. It has been a central theme of literature since the ancient Greeks.

Frankenstein tells the story of Victor Frankenstein, a wealthy young man fascinated with odd scientific theories. "It was the secrets of heaven and earth that I desired to learn; and whether it was the outward substance of things or the inner spirit of nature and the mysterious soul of man that occupied me, still my inquiries were directed to the metaphysical, or in its highest sense, the physical secrets of the world," Frankenstein relates. He has a friend named Henry Clerval who concerns himself with moral themes, "The busy stage of life, the virtues of heroes, and the actions of men."

Frankenstein has little time for such considerations. He turns to the writings of medieval alchemists, hoping to find cures for diseases but also searching for incantations able to raise ghosts or devils. Eventually he becomes fascinated by the phenomenon of life, stating, "To examine the causes of life, we must first have recourse to death. I became acquainted with the science of anatomy, but this was not sufficient; I must also observe the natural decay and corruption of the human body." After days and nights of scouring graveyards and examining decomposing corpses, Frankenstein suddenly grasps a secret: "I succeeded in discovering the cause of generation and life; nay, more, I became myself capable of bestowing animation upon lifeless matter."

He begins to assemble a huge, man-shaped being—eight feet (2.4 m) tall—that he brings to life by charging it with some sort of energy. Shelley is vague about what type, but in the introduction to a late edition of the book she mentions the experiments of Luigi Galvani (1737–98), an Italian physician who discovered that an electrical current could make the muscles of dead frogs twitch.

Instead of a having created a beauty, the Adam of a new species, Frankenstein discovers that his creation is shockingly ugly. Horrified, he flees the city and to the home of his friend Clerval, where he falls into a fever and has to be nursed for many months. Upon his recovery, he receives the news that his young brother has been murdered. Frankenstein returns home and catches a glimpse of the monster near the scene of the crime. He is convinced that his creation is responsible for the murder, and shortly afterward the two meet in a forest.

The monster forces his creator to hear his tale of hiding in the woods and watching a peasant family for months, learning to read, teaching himself to speak. His observations of people and books give him a sense of humanity and of virtue. He longs for someone to speak to. But every time he approaches people, they either retreat in terror or try to harm him. Finally he comes across a small boy in the forest. "Suddenly, as I gazed on him, an idea seized me that this little creature was unprejudiced and had lived too short a time to have imbibed a horror of deformity," the monster relates. "If, therefore, I could seize him and educate him as my companion and friend, I should not be so desolate in this peopled earth." However, the boy is equally terrified and threatens the monster—if anything happens, the boy says, his relative, Frankenstein, will exact revenge.

Hearing the name, the giant realizes that he is confronting the brother of the man who created him and murders the boy. He demands that Frankenstein construct a second creature, a bride, so that he will not be alone in his miserable existence. He has turned to violence, he claims, only because he has been abandoned by the one person who should feel a moral responsibility for him:

"I am malicious because I am miserable. Am I not shunned and hated by all mankind? . . . Shall I respect man when he condemns me? Let him live with me in the interchange of kindness, and instead of injury I would bestow every benefit upon him with tears of gratitude at his acceptance. But that cannot be; the human senses are insurmountable barriers to our union. . . . I will revenge my injuries; if I cannot inspire love, I will cause fear . . ." Out of pity and a troubled conscience, Frankenstein agrees to make the monster a wife. But later he changes his mind, fearing that he will unleash two monsters on the world. In revenge, the creature murders Henry Clerval, then Frankenstein's own young bride, and flees. Frankenstein pursues him to the North, finally reaching the Arctic, where both he and the monster die.

Frankenstein is a moral tale centered around the scientific idea that one day humans might push their power over nature too far and create organisms of their own. This idea was not entirely new. Ancient Jewish mystics and the medieval alchemists believed that life arose when inanimate material was activated by a mysterious force. This philosophy was known as *vitalism,* and during Shelley's day it was a matter of intense debate. The alternative was *materialism*—the idea that living organisms obey the same chemical and physical laws as inanimate objects. Frankenstein's vitalist method (which he refuses to tell his friends, in fear someone else might repeat his experiment) remains a secret, and the book remains a fable about what happens when people use science—or magic—to overstep certain moral limits. It echoes themes of *Faust* by Johann Wolfgang von Goethe (1749–1832), in which a scholar trades his eternal soul to the devil in exchange for ultimate wisdom and power on Earth.

This stereotype of the researcher has become fixed in people's minds and has often been used to attack scientific progress. With the rise of genetic engineering, for example, people were quick to attach names such as frankenfood and frankenfish to describe genetically modified plants and animals. The names implied that species were likely to escape control despite the best intentions of their creators. The terms are used deliberately to play on public fears. There are reasons to be concerned about where genetic science is leading, but the debate should focus on the issue at hand rather than inflammatory stereotypes. Calling a thing a monster does not necessarily make it so.

EVOLUTION, RELIGION, AND SCHOOLS

The announcement of the theory of evolution in 1858 triggered a profound change in people's thinking. Prior to that time, there was no solid scientific theory that could explain the spread of life across the Earth, the relationships between species, or their origins. While many scientists suspected that life had arisen through natural processes, the laws had yet to be discovered.

Until that happened, explanations based purely on religious beliefs seemed just as valid as any other. The work of Charles Darwin and Alfred Wallace changed that, almost immediately provoking a strong backlash from theologians and many others. While a number of religious leaders had no problem with the theory—claiming that evolution might simply have been the mechanism by which a creator put life on Earth—more fundamentalist religious thinkers took it as a direct assault on religious beliefs. The controversy continues to this day.

Initially, neither Darwin nor Wallace played much of a role in defending the theory in public. Wallace was in Southeast Asia; Darwin was plagued with ill health and occupied with family matters. His infant son died just as the first paper on evolution was to be read in public. But as critics became more vocal, some of their scientific colleagues jumped to their defense. One of the most forceful was Thomas Henry Huxley (1825–95), who later called himself Darwin's bulldog in reference to his fierce support of evolution.

Strong opposition came from figures such as Richard Owen (1804–92), the first director of the Museum of Natural History in London, and Adam Sedgwick, Darwin's former geology teacher. Sedgwick followed the hard line of the movement known as natural theology, or intelligent design, which claimed that science's only purpose was to collect evidence for the existence of a creator and bring people closer to God.

On the other hand, some religious thinkers did not see the theory as a threat, as long as it left room for a creator to have set natural laws in motion that could produce the Earth and its life. Technically, evolution made no direct statements about the origins of life: Natural selection began with the first organism that had offspring. But a number of scientists believed that even the first cell may have been produced through natural processes—some type of chemical evolution—and indeed many considered the theory eliminated the need for any supernatural explanations of nature at all. As a result, many clergymen were forcefully opposed, and they condemned Darwin in sermons, stating that the theory directly contradicted the Bible and called into question the existence of an immortal soul.

Although evolution was strictly a scientific theory, like the theory that Earth orbited the Sun, everyone attacked it on any grounds whatsoever. The first major confrontation came in a huge public meeting in Oxford in 1860 that was attended by about 700 people. Once again Darwin could not attend; he had suffered a severe relapse of a stomach illness. After delivering a long speech, Samuel Wilberforce (1805–73), bishop of Oxford and the figurehead of the antievolutionists, turned to Huxley and asked ironically whether he descended from an ape on his grandfather or grandmother's side.

Huxley could not resist making a sharp reply. Later he was quoted as saying, "If I would rather have a miserable ape for a grandfather or a man highly endowed by nature and possessed of great means and influence, and yet who employs those faculties for the mere purpose of introducing ridicule into a grave scientific discussion—I unhesitatingly affirm my preference for the ape." These probably were not his exact words, but whatever he said, the battle lines were now drawn for a major confrontation. Science and religion would be divided over the issue, and the debate would spill over into politics, education, and social philosophy.

Over the past 150 years, evolution has been attacked time and time again by religious fundamentalists, most of whom defend the belief that the Bible holds a literal account of creation and that nature bears evidence of intelligent design. This trend has been strongest in the United States. A recent tactic has been to have evolution declared only a theory, implying that it is something like a religious hypothesis that cannot be proven and thus should not be given preference over any other opinion about life. The biggest battleground has been schools; state senates have repeatedly attempted to pass laws forbidding the teaching of evolution or at least promoting a religious account of creation alongside it. Usually these movements have been motivated by Christian fundamentalist groups who have a particular interpretation of the Bible that they hope to get into the textbooks in a clear violation of the principle of separation of church and state. The laws have repeatedly been struck down by the U.S. Supreme Court.

The most famous confrontation over evolution in schools was one of the earliest: the 1925 Scopes trial ("monkey trial"), in which a substitute high school teacher was arrested and tried for breaking a state law banning the teaching of evolution in school. The trial was later dramatized in the 1955 play *Inherit the Wind,* by Jerome Lawrence and Robert Edwin Lee, and became a film several times, most notably in 1960 with Spencer Tracy and Fredric March. The play incorporates historical facts and testimony from the trial, but takes liberties with the real story.

Initially, evolution was not much of an issue in American schools—likely because most of them simply did not teach it. But at the beginning of the 20th century, things began to change, probably because of a rising general level of education and the growth of the U.S. university system. In 1900, few American pupils obtained more than a primary school education—schools had less opportunity to shape young people's ideas and opinions. By the 1930s, many more children were moving to higher grades. By that time, most scientists were strongly convinced that evolution had taken place and were including it in textbooks. This created a conflict between the research community, which expected evolution to be taught in schools like other accepted scientific principles, and fundamentalist religious groups, who saw the theory as a threat to the values they wanted their children to learn.

In the mid-1920s, legislators in several states tried to pass laws forbidding the teaching of evolution in schools. Their first success came in Tennessee in 1925, with a bill making it a crime to teach evolution or "any theory that denies the story of the Divine Creation of man as taught in the Bible."

The American Civil Liberties Union (ACLU) was alarmed because the law required science teachers to promote a particular religious view in classrooms, a violation of constitutional principles and an affront to the beliefs of the many American children who were not Christians. The ACLU wanted to challenge the law in court, but this could only be done based on a specific case. The organization placed an advertisement in a Tennessee newspaper, offering to pay all the legal expenses of a teacher willing to challenge the law. They found a volunteer in John Scopes (1901–70), a young man who had been a substitute biology teacher in Dayton, Tennessee. He was convinced by the scientific case for evolution and felt that pupils had the right to learn about it, particularly because the school's own biology book had a chapter on Darwin and natural selection. How could it be against the law to teach what was in the school's own textbook?

Dayton took on a carnival atmosphere as the town prepared for the trial. Journalists and celebrities arrived. The prosecution found a spokesman in the famous orator, fundamentalist Christian, and former presidential candidate, William Jennings Bryan (1860–1925). America's most famous trial lawyer, Clarence Darrow (1857–1938), took on the defense without charging a fee. Both felt that a great deal was at stake. The ACLU regarded the trial as a major test of individual rights versus an attempt by a religious majority to force its opinions on everyone. On one hand, democratic principles seemed to imply that the majority should be allowed to decide what was taught in schools. On the other, new scientific discoveries arose all the time, and it was felt that teachers should teach theories accepted by most scientists—even those that contradicted religious beliefs.

Darrow hoped to use the trial as a public education campaign for evolution by putting experts on biology, evolution, and even religion on the stand. The judge agreed with the prosecution, however, that the legal issue was not whether evolution was correct, but only whether Scopes had violated the law. Darrow's only option was to call Bryan to the stand as an expert on the Bible. Despite conceding that the Sun did not revolve around the Earth, Bryan claimed that the Bible should be regarded as the sole authority on matters like creation and did not need to be interpreted.

Scopes was convicted of having broken the law and was ordered to pay a fine. The conviction was later reversed on a technicality, leaving no opportunity to pursue the issue in the courts. Soon several other states in the South passed antievolution laws.

The legal issue was not addressed again until 1968, when the U.S. Supreme Court took on the case *Epperson v. Arkansas.* In their ruling, the Supreme Court stated that all antievolution laws were unconstitutional because they violated teachers' rights and represented an attempt to promote religion in public schools. Since then, fundamentalist religious groups have made many attempts to subvert ruling after ruling of the courts. Common tactics are to try to redefine religious doctrines as some type of science (as in the case of intelligent design), to portray evolution as some sort of subjective religious belief system, or to demand equal time for religious views, as if scientific theories were political campaigns.

In the meantime, the vast majority of religious thinkers throughout the world have come to terms with evolution. For example, the Catholic Church has stated that evolution and the Bible need not be incompatible—no more than the fact that the Earth goes around the Sun should challenge people's faith. In a 2006 book called *Creation and Evolution,* Pope Benedict XVI wrote that rejecting evolution in favor of faith and rejecting God in favor of science were equally absurd. His book promotes a theology of theistic evolutionism—in other words, evolution was the process by which God created life—and calls for people to stop making evolution a polarizing issue.

EUGENICS AND *BRAVE NEW WORLD*

The theory of evolution reveals that human biology is constantly changing. Humans and every other species are shaped by natural selection. In the distant future, people are likely to look, think, and behave much differently than they do today. In a few million years, the changes may be as significant as the features that distinguish humans from chimpanzees. Until the discovery of evolution, most people's thoughts about the future of their species were focused on progress. But evolution is nondirectional; it does not guarantee that people will become healthier, smarter, or better in any other way. This was disturbing to the many 19th-century thinkers who believed that human society—partly through scientific and technical progress—was improving.

Was the future of humanity completely up to chance? The arrival of genetic science suggested a way of choosing between desirable and undesirable characteristics by controlling which plants or animals breed with each other. The same thing ought to be true of humans. The genetic makeup of future generations depends on who breeds in this one. So pushing human evolution down a particular path would require a program to ensure that the right people breed with each other.

This philosophy spurred a movement called eugenics, which had two forms. Positive eugenics encouraged people with desirable characteristics to marry each other and start families; negative eugenics aimed to eliminate undesirable traits by preventing people from having children. This began with the involuntary sterilization of people in prisons and mental institutions and ended in one of the most horrifying events of history—the Holocaust, carried out against Jews and other undesirables by the Nazis. The eugenics movements that arose at the beginning of the 20th century were not responsible for the Holocaust, but they gave it a philosophy that it could drape itself in. Eugenics provided a pseudoscientific justification for racism that turned out to be naïve and based on false assumptions about how heredity works.

Ironically, two grandsons of Thomas Huxley would play a role in the rise of eugenics and the way it was regarded by society. The first was Julian Huxley (1887–1975), one of the scientist-architects who had helped bring genetics and evolution together. His brother, Aldous Huxley (1894–1963), wrote a book called *Brave New World,* which explored the social consequences of attempts to improve the human race through breeding.

Before World War II, like many scientists of his day, Julian Huxley felt that taking control of evolution would be a positive thing, comparing it to the practice of agriculture. "No one doubts the wisdom of managing the germ-plasm of agricultural stocks, so why not apply the same concept to human stocks?" He believed that social status was a sign of evolutionary fitness and that people belonged to lower classes because they were genetically inferior. From this point of view, population studies revealed a worrying trend. People of the lower classes were reproducing more than prominent members of society. If this continued, their "undesirable genes" would soon overwhelm the species. In a book called *The Uniqueness of Man,* published in 1941, Julian Huxley wrote, "The lowest strata are reproducing too fast. Therefore . . . they must not have too easy access to relief or hospital treatment lest the removal of the last check on natural selection should make it too easy for children to be produced or to survive; long unemployment should be a ground for sterilisation."

At the time, Huxley was serving as vice president of the British Eugenics Society, whose membership included a number of prominent British scientists and thinkers. The society had been founded by Charles Darwin's cousin Francis Galton (1822–1911) and headed by Darwin's son Leonard. While the emphasis of the society was mostly on positive eugenics, others had few moral qualms about the negative direction things were taking. The measures Huxley proposed in Great Britain were actually being carried out—not only in Nazi Germany, but also in the United States.

The U.S. eugenics movement arose from a long tradition of prejudice against the poor and the mentally impaired; concretely, it can be traced back to 1874, when Elisha Harris (1824–84), a physician and political reformer, became secretary of the New York Prison Association. He was a talented statistician and began to investigate a pattern he had noticed in the family names of criminals in country prisons. He traced an incredible number of "convicts, paupers, criminals, beggars, and vagrants" back to a family that had lived in Ulster County, New York, in the late 1700s. At the time of his study, six generations later, they had 623 descendants, many of whom became criminals. "In a single generation there were 17 children," Harris wrote. "Of these only three died before maturity. Of the 14 surviving, nine served an aggregate term of 50 years in the state's prisons for high crimes and the other five were frequently in jails and almshouses." In the absence of a theory of heredity, heredity was presumed to be responsible.

Another young statistician named Richard Dugsdale (1841–83) conducted a very thorough follow-up study, hoping to expose the real causes of violence. After examining records and interviewing family members, employers, police officers and many others, he concluded that the environment was more to blame in the family's tragedy than biology. What was being inherited, he wrote, was a pattern of neglect, abuse, poverty, other social factors, and physiological issues: alcoholism during pregnancies and sexually transmitted diseases that affected unborn children. He believed that the only way for the family to escape its fate would be through extensive social reforms that improved their lives, probably over two or three generations, by providing a secure environment, early education for children, and foster homes for orphans and children born out of wedlock. His conclusions went unheard. Dugsdale died at an early age, and soon after his death his work was being misused to promote the idea that criminals breed criminals. The stage was set for negative eugenics—a public campaign to improve society by ridding it of unfit members. Some of America's leading thinkers supported the movement.

David Starr Jordan (1851–1931), the first president of Stanford University, soon became one of the most outspoken figures in America's negative eugenics movement. Unlike Dugsdale, but like many scientists of the late 19th century, Jordan was convinced that heredity was far more important than the environment in shaping human behavior. He began to believe that crime and poverty were spreading like a hereditary disease, and Jordan set about ridding humanity of its degenerates. The idea took an even more negative turn when he began to associate evolutionary fitness with race and nationality. In 1907, he drew these themes together in a book called *The Human Harvest: A Study of the Decay of Races through the Survival of the Unfit.*

Jordan obtained a chairmanship within the American Breeders Association and helped change its constitution to include a platform of eugenics. The Association sponsored research into genetic studies of insanity and other mental diseases, to determine whether they could be inherited. A eugenics records office was established in the town of Cold Spring Harbor on

Long Island, New York, working closely with a nearby biology laboratory headed by Charles Benedict Davenport (1866–1944). These groups began to promote the compulsory sterilization of unfit people.

It was nothing new. In the 1890s, some physicians had begun to remove the ovaries of women with a history of psychological problems, believing that this could improve their conditions. Castration or vasectomies were performed on males as a punishment for crimes or cures for mental problems. These practices, the eugenicists said, were more humane than the death penalty, and they would have the added value of protecting society by ridding it of future criminals. Many doctors were appalled, but the practices went on.

Dr. Henry Clay Sharp (1869–1940), a prison physician in Indiana, began promoting sterilization as a solution to insanity and hereditary crime around 1900. He petitioned the governor and the state legislature to pass a mandatory sterilization law over the protests of groups of physicians who claimed the practices violated patients' rights. The Indiana legislature passed the first compulsory sterilization bill and the state's governor signed it into law in 1907. By 1930, similar laws had been passed in 30 states. By the time Sharp died in 1940, more than 35,000 people had been sterilized involuntarily in the United States.

Things started to turn around in the 1930s. Scientists like Hermann Joseph Muller (1890–1967) showed that many of the so-called studies of human heredity of Sharp and Harry Laughlin (1880–1943), a former high school principal appointed to direct a new Eugenics Record Office in New York, Laughlin ignored environmental factors such as the inequality of women and huge differences in education and health among different social classes. Muller wrote, "There is no scientific basis for the conclusion that socially lower classes, or technically less advanced races, really have a genetically inferior intellectual equipment, since the differences . . . are to be accounted for fully by the known effects of the environment." Unfortunately, these voices of reason were not heard everywhere, and as the movement declined in the United States it was on the rise in Germany and elsewhere. Only with the end of World War II and the exposure of the Holocaust did the scientific community—as well as politicians and many others—reject eugenics as a reasonable way to influence the future evolution of the species.

The British Eugenics Society had been much more hesitant to turn negative eugenics into a government sterilization program, but as the writings of Julian Huxley show, some of the members of the group were thinking along these lines before World War II. Ironically, at the same time, Julian's brother Aldous was coming to completely different conclusions about the type of society that eugenics might produce. Huxley was a novelist, and in 1932 he published his vision of the future in a book called *Brave New World.* The novel is a dark, anti-utopian dissection of a society in which the government controls not only people's biology, but also their behavior and thoughts through the use of drugs and sleep-learning.

Brave New World takes place in the year 2540—although the calendar in the future has been recalculated and in the book the date is 632 after Ford (a reference to the automaker and father of the assembly line). In this future, most of the planet is ruled by the World State, whose aim is to ensure a life of comfort and peace for its citizens. In the aftermath of World War I, H. G. Wells (1866–1946) and other authors were promoting a utopian image of the future in which a global state would create a happy, healthy society run on socialist principles. Huxley was skeptical about these rosy predictions, particularly after a trip to the United States. On the boat on the way over, he found a copy of a book by Henry Ford and began thinking about where industrialism and mass production were leading society. American society struck him as materialistic, overwhelmed by commercialism, young, sexually liberal, and focused on entertainment and pleasure. Carried to an extreme, these would be the trademarks of Huxley's nightmarish vision of the future.

The book begins in London with a tour of a reproductive factory where human embryos are being grown. Sexual reproduction has been done away with; embryos are produced by artificial insemination and raised in an artificial womb. Even a happy society needs workers to do unpleasant jobs, and science has developed a way to obtain them—a reproductive caste system. Embryos from the higher castes are unique individuals who are allowed to develop naturally. The lower castes are created by a sort of cloning, in which about 100 individuals are produced from the same egg. As they develop, chemicals are used to limit their intelligence and growth. Children are raised by the state, and various kinds of psychological conditioning are used to give them its core values. The family is replaced by a feeling of belonging to everyone.

Individualism and the desire to be alone are seen as antisocial tendencies. The masses are given a pleasure-inducing drug called soma and devote themselves to entertainment and social activities. A few areas of the world are set aside as reservations where people can live without interference from the state and have children naturally. The reservations serve as vacation spots, and sometimes nonfunctioning members of society are banished there.

The main conflict of the book involves a disillusioned psychologist from the World State named Bernard—although he is one of its few misfits, he desperately wants to fit in—and John Savage, who lives on a reservation. John is also an outcast because his mother comes from the industrial world, stranded on a reservation during a vacation. She would like to go home, and John is curious about the rest of the world. When Bernard arrives on vacation at the reservation, they accept his invitation to return to London. There, John is treated like a sideshow attraction at a circus—at first adored by all—and Bernard as a celebrity for having discovered him. But John becomes increasingly skeptical of the materialist society he has been thrust into, searching in vain for value and meaning in a world devoted to consumerism and pleasure. When his mother dies, he is torn by grief and stunned by everyone else's lack of feeling. He retreats to an isolated lighthouse to try to live alone and find peace, but crowds follow him there, hoping to be entertained. Finally, unable to cope, he commits suicide.

Huxley's novel is not a direct condemnation of genetics or biology. In 1932, DNA's role in heredity was still unknown and genetic engineering did not yet exist. But controlled breeding

The Shape of Humans to Come—the Transhumans of Patricia Piccinini

Patricia Piccinini (1965–) is a gifted Australian artist whose work includes sculptures of transhumans—mixtures of human and animal forms that are reflections on genetic engineering and the idea that people may soon have the possibility of drastically intervening in human nature. Her figures are so lifelike and thought-provoking—and sometimes so disturbing—that at least one of them was the source of a major Internet hoax.

In 2005, Arabic newspapers began printing the tale of a girl who threw a copy of the Qu'ran at her mother and was then cursed, leading to her transformation into a strange creature that was half human, half dog. The image was widely circulated in the Arabic press and became a topic of discussion in Islamic circles before it was discovered that it was a sculpture from an exhibition of Piccinini's work entitled *We Are Family*. The journalist Nizar Usman from Sudan reported, "I heard about the story from my daughter (10 years of age), she heard it in her school. Then I read it in a notice board in front of the main gate of a mosque where there was a mammoth gathering. Then I read it in *Al Hayat* daily newspaper. . . . Actually what confused people here is that there are Quranic verses saying that Allah—long ago—transferred some guilty people into monkeys and pigs."

In the catalog to one of Piccinini's exhibitions, the art critic Donna Haraway wrote, "Her visual and sculptural art is about . . . worlds full of unsettling but oddly familiar critters who turn out to be simultaneously near kin and alien colonists. Piccinini's worlds require curiosity,

(continues)

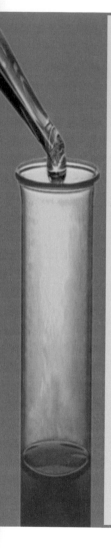

(continued)

emotional engagement, and investigation; and they do not yield to clean judgments or bottom lines—especially not about what is living or nonliving, organic or technological, promising or threatening." Haraway's essay draws connections between the history of Australia and the future as Piccinini sees it. The continent has experienced enormous shocks with the arrival of Europeans and a long, oppressive period of colonization that has seen the import of foreign species and cultures. In the future, genetic science and technology may create a similar situation across the entire world. Piccinini wants people to see what it will look like. She confronts viewers with an amazingly realistic world filled with sculptures so lifelike that they fooled newspaper editors. Her figures are disturbing hybrids of humans with wombats, humans with machines, and visions of new, unidentifiable species. Just as the original residents of Australia have had to learn to cope with change, humans will have to cope with any offspring that they decide to create.

For a 2003 exhibit in Vienna, Austria, Linda Michael described Piccinini's sculpture *The Young Family* this way: "The mother of this family lies on her side like a big sow with a litter of suckling pups. . . . Despite her status as a

practices were being used to alter crops and livestock, and there was a strong possibility that the same methods would one day be used on humans. His future society uses reproductive cloning and artificial breeding techniques, manipulating the environment that embryos develop in to influence their development. Control over society is maintained mostly through behavioral conditioning and drugs. The ultimate message seems to be that new mother, she is old. Her counterpart in the real world might be the 62-year-old woman who carried her daughter to term after a pregnancy created through IVF by Italian fertility doctor Severino Antinori. Her expression is tired, world-weary and patient, and somehow profoundly sad. She has eyes and skin with moles and hairs and veins just like us—but also the hairy back, muscular arms and hands of a primate, and a snout, long floppy ears and a tail stub. . . . It is a highly defined representation or surrogate of something. But of what?"

The species she envisions might be failed experiments from Aldous Huxley's *Brave New World*. On the other hand, a society of the future might consider them to be successful experiments. By making the figures so real, Piccinini forces viewers to meet potential artificial offspring. The works of science fiction writers and filmmakers prompt people to consider how their actions will influence the future. Piccinini's work does the same thing in a way that is somehow more subtle and intense. Of the mother figure in *The Young Family*, Michael writes, "If we see her as monstrous, is it because she threatens the continuity of our species? Does she signal the untrammeled potential of creation unleashed by transgenics? Or expose a horror of bestiality? Perhaps, in contrast, she just exposes the inevitable failure of our expectations and desires."

even if it becomes possible to manipulate human bodies and control minds toward some ideal of perfection, it will be the basic values and structure of society that will determine whether people live fulfilled lives. A society based on pleasure may not be the best future. If *Brave New World* had a single moral, it might be the old adage, Be careful what you wish for, because you just might get it.

SILENT SPRING AND THE RISE OF ENVIRONMENTAL MOVEMENTS

The 19th century ended in a wave of enthusiasm for progress. However, after witnessing chemical warfare in World War I, the atomic bomb in World War II, and the horrifying arms race that followed during the cold war, many people stopped looking at science as a solution for the world's problems and saw instead that it seemed to be contributing to them. This change in attitude was not restricted to physics. At the beginning of the 20th century, breeders and geneticists were regarded as heroes, people who might one day solve the problems of hunger and disease. Today, they are more likely to be associated with Frankenstein. This change has been influenced by the dangers evident in other branches of science—particularly concerns about the effects of chemical pollution on the environment. These burst into public consciousness with the publication of the marine biologist and nature writer Rachel Carson's *Silent Spring* in 1962, a book warning of the dangers of the long-term effects of pesticides.

One of the pesticides Carson was most concerned about was DDT, a chemical that had initially been regarded as a miracle pesticide and solution to disease. Paul Hermann Müller (1899–1965), a chemist at Geigy Pharmaceutical, created the compound. The widespread use of DDT saved millions of lives that would have been lost to typhus and malaria—a disease that was completely eliminated in many parts of the world—because it killed the insects that carried the diseases. Müller was awarded the 1948 Nobel Prize in physiology or medicine for his work.

But DDT quickly acquired a completely different reputation and along the way played a central role in the development of a large environmental movement in the United States. The reason for the birth of the environmental movement and the eventual banning of DDT was *Silent Spring,* which has been called both one of the "25 greatest science books of all time" by *Discover* magazine and received many votes as one of the "ten most harmful books of the 19th and 20th centuries" by the conservative magazine *Human Events.*

Carson's book provided scientific evidence that DDT could cause cancer and other types of environmental damage. DDT had been "sold" to the public in extensive publicity and advertising campaigns in which its virtues were emphasized—had potential dangers been covered up? A lone woman was saying that it was absorbed by crops and entered the diets of animals and humans. Her data suggested that it had a negative effect on birds, fish, trees, and many other forms of life.

Rachel Carson's book *Silent Spring* gave a scientific account of the effects of DDT on the environment, woke up the American public to the dangers of pollution, and helped spawn the modern environmentalist movements. *(NOAA/Department of Commerce)*

Carson immediately found herself in the middle of a public relations war in which researchers ridiculed her science and the chemical industry attacked her personally; she was labeled hysterical, and industry spokesman Robert White-Stevens said, "If man were to follow the teachings of Miss Carson, we would return to the Dark Ages, and the insects and diseases and vermin would once again inherit the earth." Most of the attacks were grossly unfair. Carson had never proposed banning all pesticides, and she was well aware of their importance in fighting disease. In *Silent Spring* she wrote: "No responsible person contends that insect-borne disease should be ignored. The question that has now urgently presented itself is whether it is either wise or responsible to attack the problem by methods that are rapidly making it worse." She pointed out that the overuse of pesticides had produced insects that were resistant to them: "The list of resistant species now includes practically all of the insect groups of medical importance. . . . Malaria programmes are threatened by resistance among mosquitoes. . . . Practical advice should be 'Spray as little as you possibly can' rather than 'Spray to the limit of your capacity.' . . . Pressure on the pest population should always be as slight as possible."

This was in good keeping with evolutionary science: A challenge such as a pesticide can suddenly put enormous pressure on a species. If chance provided some mosquitoes, for example, with genes that offered them partial resistance to a pesticide, they would likely undergo rapid positive selection, and within a short time the genes would spread through the population. Soon farmers would be back at the same place they started. It was an important lesson that would need to be learned over and over again. For example, it took doctors decades to learn that the overprescription of antibiotics was promoting the evolution of highly resistant bacteria. Such cells would have arisen anyway—bacteria reproduce so quickly and undergo so many mutations that they are amazingly adaptable—but the overuse of drugs has cleared a path for resistance genes to spread at an alarming rate. It increases the chances that a person who desperately needs antibiotics could be infected with a strain that does not respond to them.

Silent Spring was well written and controversial, which made it so popular that many people became alarmed. Its publication put pressure on the government, the chemical industry, and scientists. A committee was appointed by President John F. Kennedy to look into the matter. Its findings supported those of Carson and led to changes in the way pesticides were regulated by the government. This planted the seeds for a strong American environmental movement that turned its scrutiny on pollution and chemical waste products, many of which turned out to cause cancer.

From the beginning, the issue had a polarizing effect. Environmentalists felt that they could not entirely trust scientists, the chemical industry (who might be paying them), or government regulations to protect citizens from the products of research. The result was a conflict that continues to the present day. It can be seen in the multitudes of environmental partners as well as platforms of the green parties in the United States and many other countries. Environmental concerns blend themes of consumer advocacy and quality control, skepticism of industrialization and global business practices, concern for the environment and endangered species, and fears that meddling with nature on the part of scientists will lead to disaster. Thus when genetic engineering came on the scene, the stage was set for a confrontation.

GENETICALLY MODIFIED CROPS AND THE MARKETING OF SCIENCE

Genetic engineering arose in the late 1970s and early 1980s against a backdrop of new public concerns about technology and the environment. By the time the first genetically modified foods were put on the market in the 1990s, nearly every day's news brought reports of some new, cancer-causing substance. In other words, it was dangerous to release the products of science into the environment. Industrial chemicals and pollution had caused a huge hole in the ozone layer over Antarctica, allowing dangerous solar radiation to penetrate the atmosphere and increasing people's risk of skin cancer. Atmospheric studies warned of a greenhouse effect that could dramatically change the global climate. Asthma, allergies, and cancer were on the rise. Governments began to pass laws to minimize environmental pollution and to ban cancer-causing substances.

From the beginning, scientists were interested in using genetic engineering to improve crops. One reason was the concern that traditional methods of food production could not keep up with the number of mouths to feed. At the World Food Summit in Rome in 1996, experts stated that the world would have to double its production of food within the next 30 years just to keep pace with population growth. It has been estimated that 800 million people on the planet currently suffer from malnutrition and starvation. A solution might be to create *genetically modified organisms* (GMO) by directly modifying the genes of crops. As well as improving their size, taste, shelf life, or nutritional value, foreign genes can offer protection from insects, fungi, and other parasites without the dangerous side effects of pesticides. Members of the growing ecological and environmental movements protested that genetic engineering might upset delicate balances in nature.

One issue that makes the debate so complex is that so many of the participants have political, financial, or ideological agendas. Companies interested in creating new foods—citing humanitarian reasons and the hope of making profits—claimed that farming had always produced highly artificial crops and the products of genetic engineering would be no different in any significant way. Critics said that artificially modified organisms would automatically have an unnatural advantage over plants bred by other methods and that introducing them into the environment in huge quantities would change the way genes normally move through a population.

The first food brought to market was the Flavr Savr tomato, by a California-based biotech company called Calgene, in 1994. Research had shown that a protein called polygalacturonase played an important role in how tomatoes rot because it softened cell walls as the fruit ripens. Inserting a second gene that interfered with the protein yielded tomatoes that could be stored longer without losing their taste.

The U.S. Food and Drug Administration (FDA) examined the plant, deciding that it did not pose a health hazard to people and could be put onto the market without special labeling. Although customers in the United States and Europe were initially enthusiastic, in the long run Flavr Savr was not competitive. It lost out to other long-lasting, non-GMO brands that customers preferred—Calgene had not used the best strain of tomato to begin with, and the company had little experience in growing and marketing foods. But the GMO era had begun, and the following year Calgene was bought by the company Monsanto, which has become a major producer of many types of genetically modified foods.

In Europe, public acceptance of GMOs quickly plummeted as consumers became concerned that there might be unknown risks. Protesters demanded strict governmental controls (such as bans on imports, or at least clear labels marking food as a product of genetic engineering). The change in attitude was partly due to outbreaks of a deadly disease called bovine spongiform encephalopathy, or mad cow disease, which was caused by cows, normally herbivores, being fed the remains of other cattle. The disease was then passed along to humans who ate the animals. Once again, political and business agendas came to bear as governments initially shrugged off the threat. By the time their attitudes changed, many Europeans had lost confidence in governments' ability to regulate foods.

Tomatoes were quickly followed by genetically modified soybeans, cotton, and maize. Some of the new varieties improved the nutritional value of foods that are the core of people's diets in many parts of the world. Corn and rice lack vitamin A, which is essential to the development of the eye. Adding genes to these staple crops has helped reduce blindness and other symptoms of malnutrition that have plagued children throughout the world. Plants have also been made resistant to herbicides so that weeds can be killed without damaging crops. Tomatoes, cotton, corn, and many other crops fall prey to caterpillars; researchers have added a natural toxin, a protein called Bt from the bacteria *Bacillus thuringienses,* that kills the insects. Sweet potatoes in Africa have been made immune to viruses. Changes in species of rice have produced strains that can survive floods, and other plants have been modified to tolerate high levels of salts or acids in the soil.

The number of GMO crops continues to increase dramatically, particularly in the United States, Argentina, Canada, and China. Recently it has been estimated that about 75 percent of foods on the shelves of stores in the United States contain at least one modified ingredient. In other countries, the trend has grown at a somewhat slower pace, but overall GMO crops are winning an increasing share of the world food market. By 2005, approximately 60 percent of the world's soybean fields, 28 percent of the cotton, and 14 percent of the maize were devoted to these crops.

Decisions to develop and grow GMO foods are based on the profit they are expected to bring, as well as other motives. Businesses have sometimes engaged in questionable practices to gain an advantage over their competitors to the detriment of farmers and economies in developing nations. The practices have also raised new legal issues such as questions of ownership. The creation of a new crop requires a huge investment in basic research, laboratory experiments, costs of growing, and risk assessments. Companies need to recapture these expenses through profits, which are best ensured by maintaining ownership of their crops.

There has been a growing interest in the production of genetically modified animals for foods as well, but the efforts have met with technical, ethical, and legal challenges. It is much more difficult to develop these animals than plants. Often, a new plant can be grown from an existing one simply by taking a single cell. In animals, new genetic material can be introduced into the very early embryo so that the animal's egg or sperm cells contain the gene. The methods are not perfect, and many generations may be needed to obtain a strain with the gene.

Other efforts are underway to create pigs that produce leaner meat and to use animals as factories for drugs like insulin. The same strategy has been used to make another hormone called erythropoietin, which stimulates the development of red blood cells, and is used as a treatment in anemia and some forms of kidney disease.

Another use of genetic engineering is to create drugs that might be delivered to people through foods: chickens whose eggs contain antibodies or bananas containing vaccines. Human genes have been added to animals so that they produce the human forms of proteins, sometimes in their milk. The goal is not to produce molecular "cocktails" to deliver therapeutic genes in a drink—but to produce them in a form that can easily be extracted from an animal (by milking it) and then purified.

A major concern for the public has been the fear that genetically modified plants and animals might have an unforeseeable impact on the environment or their bodies. Here too some precedents had shaped public attitudes about introducing new species into the ecosphere. In the 1960s, scientists brought a fish called the Nile perch into Lake Victoria in Central Africa, the source of the Nile River. The perch was so well suited to its new home and multiplied so quickly that native species of fish have been virtually wiped out. The story is told in a 2005 documentary called *Darwin's Nightmare,* written and directed by the Austrian filmmaker Hubert Sauper. The film explores the human and economic impact of the fish, now such an important source of food that it is bartered to Russian buyers in exchange for weapons. Other cases of transplantation have been unintentional, such as the transport of pests in food containers or small animals such as snails that are often carried along in the ballast water of ships. When ships travel across the world and illegally empty their tanks, the result may be infestations or the disruption of local food chains. (This has been going on as long as people have traveled the globe, carrying along seeds, animals, and hitchhiking parasites—so today's normal environment must also be seen as the product of steady contamination.)

These situations are not directly related to genetic engineering, but they have set the backdrop for people's responses. While some of the arguments against genetically modified foods or other organisms are based on scare tactics, rather than realistic estimations of risk, scientists admit that the effects of these foods or other organisms are impossible to predict with absolute certainty. Every organism lives in a complex network of interactions with every other, from bacteria in the soil to other plants and animals. Testing a new strain's effects on all of them would be impossible.

Legally, however, the question became whether genetically modified species should have to meet far stricter standards of safety than any other new product brought onto the market. Many people thought that they should because they felt that genetic engineering was tampering with nature. Defenders of GMOs point out that farming also alters organisms through selection; it works with the changes in plants and animals that arise through mutations and other natural processes. These changes occur in random genes and are completely unpredictable; they may have equally strong effects on the environment, but these effects may never be studied under controlled conditions in the laboratory.

In the United States, the FDA is responsible for approving modified foods intended for market after extensive testing in laboratories. The Flavr Savr tomato was approved after the FDA determined that it did not pose a health hazard. Producers were not required to give it special labeling. In 2003, a survey conducted by ABC News showed that 92 percent of Americans believed that "the federal government should . . . require labels on food saying whether or not it has been genetically modified or bio-engineered." The percentage had steadily risen since similar surveys in 1998 (82 percent in favor of labeling) and 2000 (86 percent).

Some common fears, such as the idea that modified genes from a plant might enter a person's body and cause health problems, are simple misunderstandings about how genes work. Scientists have never discovered a case where a gene from a plant has been taken up by the human genome by eating; all food contains foreign DNA, and it is destroyed during digestion. (Some of the concerns may have arisen because of mad cow disease, but there the cause is a protein fragment; genes would not behave the same way.)

While most scientists admit that it is impossible to calculate all the risks involved in the creation and spread of genetically modified organisms, they are concerned that the debate has not been balanced and sufficiently informed by facts. Many feel that science fiction movies, negative publicity, and misunderstandings have given the public a false idea of what GMOs are and how risky their use might be. Just as the media present much more bad news than good, problems with GMOs receive far more attention than the positive effects they have had on millions of people's lives. GMOs have been designed to create new foods, curb hunger and starvation, and prevent disease. Not using science to try to solve some of these very grave problems—when there is no evidence that these crops pose a greater threat than the products of traditional agriculture—would be ethically very questionable. Discussions should not be entirely focused on risks; they must also weigh the potential benefits and give equal consideration to the consequences of not taking action at all.

Genetically modified corn growing in a field in Europe. In recent years, Europeans have shown a growing resistance to the planting of these crops; in 2009, a popular brand from the company Monsanto was outlawed in Germany.

Some studies have shown that GMOs do occasionally have unintended consequences on the environment. Cotton bearing the bacterial toxin Bt was put onto the market in 1996 by the company Monsanto, a major developer of GM crops, in order to ward off a type of moth larva called the bollworm. This pest puts such a huge dent into farmers' yields that growers in China, India, and the United States plan to start planting Monsanto cotton, despite the fact that seeds cost three times as much as other plant strains. In the meantime, more than one-third of the cotton grown across the world has the Bt gene. In the United States, more than 70 percent of cotton crops have been genetically modified.

Initially, this saved farmers money because they could cut back on the amount of costly pesticides needed to protect the crops—their use dropped by more than 70 percent. But a study completed in 2006 by researchers from Cornell showed that within seven years, Chinese farmers were using just as many pesticides as before. Not because the bollworm had evolved resistance—but because in the absence of the larva, other pests such as leaf bugs were moving in. Those, too, have to be combatted using pesticides.

These facts have rightly been cited as examples of the unintended problems that can arise through the use of GMOs. Yet it is important to remember that bollworms are not really natural pests that are being chased out by unnatural ones. The vast cotton fields that they infect are not natural; the plant itself is the product of millenia of artificial breeding practices. Farmers in the Indus Valley (located in today's Pakistan) began cultivating it from a wild plant more than 6,000 years ago. Within 1,000 years, another form of the plant was being cultivated in Mexico. Today the largest fields are found in West Texas. They do not grow there naturally, which means that the bollworms that infect them are not more natural than secondary pests such as leaf bugs. Growing cotton also requires a great deal of water that has to be diverted from other sources, so another aim of breeding and genetic engineering is to reduce the amount of water that the plants need to thrive.

Cotton is a good example of the complexity of this issue. Inserting the Bt gene is not the only way researchers are modifying the crop. Cottonseeds cannot be eaten because the plant produces a toxic substance called gossypol—which also acts as a natural pesticide, because very few organisms can digest it. However, scientists at Texas A&M University have used genetic engineering to produce a strain of the plant whose seeds do not contain gossypol. This may make it possible to turn cottonseed into a source of food for livestock and even humans. That would be of immense importance because cotton is already one of the most-cultivated plants in the world.

These factors point out some major social issues concerning GMOs that has little to do with science—they are also commercial products. Businesses, governments, and others have invested heavily in their development and obviously have an interest in turning crops into profit. Like any application of science, whether they ultimately have a positive or negative influence on society will depend on how wisely they are used by those who develop them, those who sell them, and ultimately the consumers who decide to buy them. The same types of issues surround other potential uses of genetic engineering that are explored in chapter 4.

CLONED DINOSAURS AND INVADERS FROM SPACE

Frankenstein, Brave New World, and many other science fiction novels established one of the most prominent themes of the genre: Bad things happen when scientists overextend their reach, trying to manipulate things that they do not completely control, using technology that they do not fully understand. These were prominent themes in the fiction of Michael Crichton (1942–2008), a trained physician who became one of the most popular authors of the late 20th and early 21st centuries. After writing a number of thrillers, many of which somehow integrated technology or scientific themes, Crichton achieved his breakthrough with a 1969 novel called *The Andromeda Strain.* The plot of *The Andromeda Strain* centers around a team of biologists fighting a deadly microorganism that seems to have come from outer space, brought to Earth aboard a satellite. Researchers retrieve the satellite from the small town of Piedmont, Arizona, near its crash site. It has been opened by curious citizens, unleashing a plague that wipes out everyone in the town except for an old man and a small baby. The team takes the satellite and the survivors to an underground laboratory in Nevada, where they hope to isolate the microbe and find a cure. The facility is equipped with an ultimate fail-safe measure—a nuclear weapon that will destroy it (and everyone inside) in case the organism escapes.

Throughout the book, scientists are confronted with situations that arise because they assume they are in control of things. But nature is messy, chaotic, and complex, and unforeseen events nearly lead to disaster. Problems range from the trivial—a piece of paper gets stuck in a teletype machine, preventing the transmission of a vital message—to mistaken assumptions that have

A predominant theme in many of Michael Crichton's science fiction novels is how the complexity of living systems brings near disaster to scientists who think they have nature under control.

nearly fatal consequences. Having no previous experience with extraterrestrial life, the researchers assume that the radiation from an atomic blast will kill it. By the time they discover that this would have the opposite effect—causing the organism to absorb energy, mutate rapidly, and become infinitely more dangerous—it is almost too late. Ironically, the primitive technological failure and the huge scientific mistake cancel each other out. At the climax of the novel, the bomb must be stopped, but a design flaw in the facility makes this almost impossible. The world is saved not by brilliance, but by a combination of ingenuity and good luck.

Most of Crichton's later books are built around stories with similar elements. In his 1990 novel *Jurassic Park,* a wealthy biotech billionaire named John Hammond has found a way to create living dinosaurs, using DNA found in insects that drank dinosaur blood and then became preserved in amber. The DNA is not intact, but researchers have filled in the gaps with genes from modern relatives, such as birds and amphibians. The information has been used to create complete, artificial genomes of several dinosaur species. They are inserted into cells, transplanted into eggs, and cloned. The animals are raised and kept on an island called Isla Nublar, which Hammond intends to turn into a new entertainment park.

Because of the obvious dangers of dinosaurs, various safeguards have been put into place to maintain control. To prevent the dinosaurs from reproducing, they have all been engineered to be female. They are meticulously tracked, and every animal is accounted for. But the book opens with signs that things are not going as smoothly as planned. Tourists have been attacked by an unidentified beast in nearby Costa Rica. The billionaire brings a team of experts to the island to investigate, hoping for a clean bill of health. The team soon discovers that nature has found a way to overcome man's limitations; the dinosaurs are having offspring. One by one, the island's security systems fail. As the group's mathematician has predicted from the very beginning, complex systems—like organisms—can never be completely understood and controlled.

Jurassic Park is a good example of how scientists are portrayed in Crichton's works. They are both heroes and antiheroes, human beings who suffer from the same weaknesses as everyone else. The crisis in the novel is partly the fault of researchers and their attempts to use gene technology to an end that is far too ambitious. The problem is compounded by human error, the greed of individuals and companies who are trying to profit from scientists' work, mistakes in planning, and flaws in technology. The situation leads to a rapid breakdown of control on the island, and the humans become prey of a dinosaur population that is completely out of control. As in *The Andromeda Strain,* however, things do not end nearly as badly as they might. A few people manage to escape the island, through a mix of knowledge and luck. Supposedly the dinosaurs are destroyed—only to return in a 1995 sequel called *The Lost World.*

Crichton's 2004 *State of Fear* takes the relationship between fiction and science further, taking aim at the politics of global warming. The antagonists are a group of ecoterrorists who are staging a number of natural disasters (actually of their own making) that will cost a huge number of lives; their aim is to convince the world that the environment is rapidly being destroyed. As he wrote the novel, Crichton believed that there was not enough scientific evidence to support some of the dire predictions that many scientists were making, and that there was little evidence that the solutions proposed would actually have an effect on the problem.

State of Fear was heavily criticized by a number of experts on global warming, journalists, and environmentalists—including many whose papers Crichton had cited in the footnotes. They claimed that Crichton had misinterpreted and misused their findings. In an article in the January 20, 2005, issue of *Nature*, Myles Allen of the University of Oxford's Climate Dynamics Group wrote, "Although this is a work of fiction, Crichton's use of footnotes and appendices is clearly intended to give an impression of scientific authority."

The history of science fiction shows that visions of the future can be useful and help people consider the ethical implications of research in fiction, before they have to be confronted in the real world. But as Allen and many others point out, the distinction needs to be preserved. A hallmark of the early 21st century is a rapid blurring of lines between science and fiction, science and politics, entertainment, news, and many other types of reality and fiction. Novelists have always taken liberties with scientific facts in an attempt to create art, promote personal points of view, and sway people's opinions. This has taken on a new dimension at a time when the technology of the entertainment industry is able to create compelling images that often seem more real than the real world. It is easy and dangerous for people to become confused and mistake fictional risks for real ones. If people can no longer recognize the difference between real science and entertainment, their opinions will be at the mercy of those who plan next week's television schedule and who are best at using the media in support of ideological agendas. And then the sort of brave new world imagined by Huxley and other anti-utopians will likely not be far behind.

3

Studying Life in the Post-Genome Era

Modern molecular biology began with the discovery of the structure of DNA in the early 1950s. Over the two decades that followed, scientists worked out the way that cells use the information in the genome: by building *messenger RNA* molecules (mRNAs) based on the sequences of genes, and then using the mRNAs to make proteins. Once the roles and relationships of these molecules had been clarified, the main aim became to understand the function of each gene in the life of a cell and organism.

Everything that happens in the cell involves complex networks of dozens or hundreds of molecules. Until the 1990s, however, technological limitations meant that scientists could usually investigate the functions one by one; at best, they could observe the behavior of a few molecules at a time. Now this has changed thanks to rapid developments in biotechnology. At the dawn of the 21st century, which many researchers call the beginning of the post-genome era, the strands of biology, physics, medicine, and modern disciplines such as biocomputing have come together to produce a rich, modern view of life. This chapter introduces some of the most important, cutting-edge methods in biology and describes how they have begun to change our view of life.

GENOTYPES AND PHENOTYPES

A huge amount has been learned about life since the birth of molecular biology in the mid-20th century, but some of the most interesting questions have yet to be answered. A main goal of current research is to understand the relationship between an organism's *genotype*—the complete set of hereditary information in its genome—and its *phenotype,* which means the structure of its body and anything else about it that can be observed. The phenotype includes tiny features such as the molecules present in its cells, as well as much larger phenomena such as details of the stages of an organism's development or its behavior. The genotype is invisible—except in the sense that it can be read by obtaining a DNA sequence—and some of the information it contains may never appear in the phenotype. For example, a woman may have inherited a form of a gene that causes color blindness and pass it along to her sons without herself becoming color-blind.

Ideally, scientists would like to be able to look at a genome sequence and make detailed predictions about the phenotype it can produce. This is already possible to a certain extent, in situations such as the following.

- Matching a fragmentary DNA sequence to a particular species or its place in the living world. All organisms on Earth have related genes that have undergone mutations as species evolve. Each species inherits the "spelling" of its genes from its direct ancestor; subsequently they undergo new changes. This chain of events can be read from DNA sequences, so an organic sample can be used to identify the species from which it comes. If the species is unknown, it can usually be classified into a *clade* (a group of organisms that belong to the same evolutionary branch). The early summer of 2005 saw an amusing application of this principle when a number of people reported seeing a large, unidentified creature near the town of Teslin, in Yukon, Canada. Media reports suggested the animal might be bigfoot—an unidentified, ape-like creature—that some people believe inhabit sparsely populated areas of North America and other regions of the world. Strands of unusual hair were found in the vicinity of one of the sightings. The hair was given to

David Coltman, a geneticist at the University of Alberta, for analysis. Coltman obtained DNA from the sample, sequenced it, and compared it to other known species. The DNA turned out to have come from an American bison. Coltman told a bigfoot enthusiast that he believed the sample had come from a rug made of bison hide and had probably been deliberately planted.

- Predicting that a plant or animal will have specific features based on an analysis of its alleles. Mendel showed that greenness is dominant in peas, so a plant will have green seeds even if it also carries a recessive allele for yellowness. Therefore, if scientists discover the greenness allele in a pea plant, they can predict the color of the seeds it will produce. This is also true of other *monogenic traits* where one form of a gene is dominant. But only a few traits in humans are truly monogenic. One is having a cleft chin; another is the ability to roll the tongue into a U-shape. Most characteristics, such as skin color, are the result of contributions from many genes. In those cases, it is much more difficult (and often impossible) to make an accurate prediction of the characteristics a person will develop based on an analysis of his or her DNA.

- Predicting that an organism will develop, at some point, certain diseases. Researchers have connected thousands of alleles to monogenic diseases. For example, a person born with a particular form of a gene called huntingtin is virtually certain to develop Huntington's disease. The defect causes the loss of cells called medium spiny neurons, which play an important role in coordinating movement. The problem is usually discovered when a person begins to experience uncontrollable, jerky movements that come from this loss of coordination. Other diseases are thought to be caused by combinations of specific alleles. These are much more difficult to identify. Many seem to be more susceptible to environmental influences than monogenic conditions, so it may be impossible to predict with accuracy whether a particular individual will develop the disease. There are often big differences in the degree to which people who have inherited such a multifactorial trait develop symptoms. Even interpreting the results of a test for a monogenic disease has to be done by experts with care, because in some cases a second gene might be able to reverse the effects of a defective one.

• Reconstructing ancient forms of genes and features of organisms that no longer exist. Two species usually have similar features because they inherited the characteristics from their common ancestor. There are exceptions. Fangs evolved many times—snakes and cats did not inherit them from the same ancestor. Wings evolved separately in birds and bats. But in general the principle holds, and it has been used to make hypotheses about shared ancestors. For example, scientists currently estimate that chimpanzees and humans descend from a primate that lived between 4.5 and 6.5 million years ago. Fossils of this ancestor have not yet been found, but a comparison of the two species gives some good hints about what it must have been like. Researchers possess more information about its genome than its appearance or behavior. Overall, 96 percent of the two genomes are identical, and many neighborhoods within the genome, containing genes and other information, are about 99 percent identical. This allows scientists to reconstruct nearly the entire genome of the ancestor. If they can learn more about how genes interact to build a body, they may be able to construct a very accurate model of its biology and appearance.

All of these types of work would be given a boost by a detailed understanding of the connection between genotypes and phenotypes, especially the early diagnosis of genetic diseases. In 1990, Mary-Claire King (1946–   ), professor of genetics and epidemiology at the University of California, Berkeley, discovered that some alleles of a gene called BRCA1 are strongly correlated with a much-increased risk that a woman will develop breast cancer. At the time, the idea that there might be a genetic basis for cancer was very controversial. Further research has shown that she was right and has also offered a partial explanation. The function of BRCA1 is normally to help repair damage to DNA, which can occur as a result of exposure to radiation, other types of environmental contamination, or through mistakes in cell division. A healthy form of BRCA1 can step in and repair the damage, which might otherwise trigger cancer. For this reason BRCA1 and similar molecules are called *tumor suppressor genes.*

Mary-Claire King, professor of genetics and epidemiology at the University of California, Berkeley, discovered that some forms of the BRCA1 gene inherited by women are strongly correlated with a high risk that they will develop breast cancer. *(Peter and Patricia Gruber Foundation)*

If BRCA1 breaks down, it no longer does its job, and cancer-causing defects can slip through. However, not all women who have inherited the defect develop tumors. Cancer often requires other random, unpredictable mutations in genes—either through natural mistakes as DNA is copied or influences from the environment—and a woman might be lucky enough to avoid them. Finding a way to calculate risks more accurately would be very helpful both to patients and doctors as they face difficult decisions about therapies.

GENOMIC TECHNOLOGIES

In the post-genome era, the molecule-by-molecule approach to studying life has given way to technologies that can observe the activity of the entire set of genes in a cell and throughout an organism's body. One aim of these experiments is to get a better idea of how molecules participate in cellular processes. There are also more ambitious goals, such as discovering the causes of cancer and other diseases, finding better ways to diagnose them, and learning why different people respond to the same medication in different ways. This information will be an important step on the road to personalized medicine, in which details of a person's individual genome are taken into account while diagnosing tendencies to diseases and designing treatments.

These methods will be the most powerful when researchers and doctors have access to the complete genetic codes of individuals. (The human genome sequence that has been obtained now is a reference version, obtained by combining the DNA of several anonymous individuals; no one person has exactly this sequence.) Realistically, this will only happen if the cost of obtaining a complete sequence drops to a reasonable level. The U.S. National Institutes of Health (NIH) has launched a program whose target is to bring the price to $1,000 dollars by the year 2014. If that target can be met—in 2014 or the near future—it will be feasible to sequence individual genomes on a large scale, and it will signal the beginning of an era of personalized genomics. How far the information affects a person's medical care will depend on what has been learned about the genetic causes of disease.

Currently, researchers have linked about 6,000 single genes to diseases. Multiple genes are thought to contribute to many more illnesses, but finding them requires studies of very large families. If a parent has a disease caused by two genes, only a fourth of his or her children will inherit both of them. If the disease is recessive, then both parents have to have both genes—and only one in every 16 children may display symptoms. Not many families have that many children, so the disease might not appear for several generations. If it does, it may not be recognized as a hereditary problem. A similar situation confronts researchers who wish to understand very complex interactions between organisms and the environment, such as the links between diet and disease. There are so many variables in the environment that isolating the important ones might be impossible without collecting huge amounts of data.

Genomic technologies will provide a shortcut in diagnosing known genetic problems. In early 2009, nearly 2,000 types of genetic tests were available, but most target a specific disease (or just a few). With access to a person's complete genome, on the other hand, the sequence can be scanned by computer, for all known disease markers.

Most of the technologies that monitor the activity of the genome do so by detecting RNAs, which are produced when a gene is switched on, or proteins, which are made using the information in messenger RNAs. The main methods are described below.

- *DNA microarrays* (DNA chips) use probes made of DNA attached to glass or another material to detect RNA molecules. DNA and RNA are made of the same basic subunits—nucleotides—that bind to each other if they have complementary sequences. One common type of probe contains sequences that are complementary to RNAs for every human gene. RNAs are extracted from cells, and if the sample contains an RNA made from a particular gene, it binds to the probe and emits a fluorescent signal. Microarray experiments usually compare two types of cells, such as healthy human cells and those taken from a tumor. The readout of the experiment shows which genes are more active in one cell than the other and which behave the same way. If a gene is more active in one type of cell, this may show that it plays an important role in the process that is being studied—for example, the development of a tumor. The method was developed in 1994 by Pat Brown, a biochemist at Stanford University and the California-based company Affymetrix, and is now used almost universally in laboratories throughout the world.
- *Tiling arrays* are specialized micro-arrays that are often used to look at the behavior of regions of the genome that do not contain genes—in humans, this material accounts for at least 98 percent of the sequence. Until recently, the function of most of this DNA was a complete

Readouts of DNA chips like this one show researchers how gene activity varies in different types of cells. Comparing a cancer cell to a similar but healthy one, for example, reveals genes that play a role in the development of tumors. *(WormBook)*

mystery—it was often called junk. Many researchers supposed that it was simply extra material that had accumulated over the course of evolution, such as artifacts of ancient genes that had undergone mutations and become nonfunctional. Now scientists have discovered that cells transcribe an enormous amount of this material. Often the products are small RNA molecules called *microRNAs*. They are not used to make proteins; instead, many of them dock onto other RNA molecules with complementary sequences and prevent their translation into proteins. Tiling arrays usually consist of very short probes to which small RNA molecules bind. In 2004, Jason Johnson and Eric Schadt used this method to survey the complete gene activity of two human chromosomes. Making the array was a huge task involving the construction of more than 3,700,000 individual probes. The results were stunning, revealing that cells made about 3,000 RNAs unconnected to any known gene—just on the two chromosomes. Schadt and his colleagues concluded that about one-fourth of these might represent real genes that had gone undetected. In 2005, Jill Cheng, Philipp Kapranov, and their colleagues at the company Affymetrix put together a tiling array for 10 human chromosomes, containing more than 74 million probes, covering about 30 percent of the human genome. They discovered that on the average, about 10 percent of each chromosome is transcribed into RNA. This is between five and 10 times the amount of RNA known to encode proteins.

- *Chromatin immunoprecipitation* (ChIP) surveys the genome to discover where particular proteins bind to DNA. Usually the activation of a gene begins when a protein called a transcription factor binds to a sequence. But a single transcription factor can usually activate many genes, and it has been difficult to discover all the targets of these molecules. The first step in the method is to fix proteins to DNA. Then enzymes are used to chop combinations of DNA and proteins into fragments. Antibodies are used to extract specific proteins and the DNA that they are bound to. ChIP is often used in combination with microarrays to identify the target DNA sequences.

- Mass spectrometry, described in chapter 1, identifies the proteins that are present in a sample taken from a cell. One of the most interesting uses of the method is to analyze the composition of protein machines. Most proteins carry out their jobs in complexes ranging from a few up to 100 molecules—machines which are continually dismantled and rebuilt to perform different functions. This activity is central to understanding the cell, but it has been extremely difficult to observe. In 2002, researchers in Heidelberg, Germany, discovered how truly dynamic this situation is. Anne-Claude Gavin and Giulio Superti-Furga of the young biotech company Cellzome worked with scientists at the European Molecular Biology Laboratory to capture the first complete view of the machines at work in a yeast cell. They used a new method to extract whole machines from cells and analyzed their components with mass spectrometry. They found that 17,000 proteins form at least 232 machines. Many of them work in a snap-on way; they have a core of preassembled pieces and when it comes time to do a certain job a few more are added on. A machine may remain inactive until that happens. This gives the cell a way to control its activity. To be switched on, it may need to borrow the missing pieces from other protein complexes, or components may have to be made anew.

- Live cell arrays are slides made of glass, silicon, or another material on which living cells are grown in different compartments. Molecules or substances are introduced into the compartments to study their effects on the cells. One common use is to watch how cells respond to a drug or a toxin. Another type of experiment exposes the cells to various microRNAs, which block the production of specific proteins. This often reveals the functions of the molecules. If the loss of a protein disrupts the cell cycle and makes cells divide too often, it may be a sign that the molecule plays a role in cancer.

Each of these methods reveals a slightly different aspect of the complete molecular activity that takes place in a cell. In combination, they are giving researchers a new look at how information in the genome guides the life of an organism.

BIODIVERSITY AND METAGENOMICS

How many types of viruses and bacteria live in a person's body or the soil of a farm? What effects do pesticides, genetically modified crops, or the transplantation of organisms to new regions have on the environment? How great is the effect of global warming and human overpopulation on food chains across the globe? Answering these questions will depend on our ability to measure *biodiversity*—a survey of all the organisms in a particular environment or on the Earth as a whole. Even without considering most microbes, scientists have already identified and named about 1.6 million species (more than half of which are insects), but they estimate that the Earth holds many times that number. Some researchers estimate the number of insect species alone at 10 to 30 million. Few scientists are willing to venture a guess as to the number of types of bacteria and viruses that exist; it is sure to be many, many times more. Recently, the arrival of rapid DNA sequencing methods and databases of known sequences have given researchers their first deep look at this invisible world, which humans, plants, animals, and fungi are heavily dependent on.

The new approach, called *metagenomics,* was conceived by Norman Pace, a molecular biologist now at the University of Colorado at Boulder, in 1985. Up to that time, DNA sequencing efforts had focused on humans, important laboratory organisms such as flies and mice, and microbes that had been cultured in the laboratory. The idea was to start with one species, obtain its complete sequence, then move on to the next. Pace wondered what would be found if he simply sequenced all the DNA in a sample of water or soil—more like the genome of a global positioning system (GPS) coordinate.

This was particularly important, he felt, because most microbes could not be raised in laboratory cultures. In nature, they usually live in complex communities in which thousands of different types depend on each other for survival. These living networks fulfill vital functions for humans and the ecosphere. They lie at the base of every food chain, and they play a crucial role in regulating the chemistry of the atmosphere and the water supply. Every liter of ocean water, for example, holds billions of cells that help plants remove carbon dioxide (CO_2) from the atmosphere. But scientists have likely only seen a small fraction of them and have little idea of their roles in supporting other kinds of life, including humans.

In 2002, Mya Breitbart and Forest Rohwer of San Diego State University began taking an in-depth look at the ocean using a metagenomics approach. They discovered that 200 liters of seawater contain DNA that comes from more than 5,000 species of viruses, and one kilogram of marine sediments may contain up to one million species. Samples of human feces contain more than 1,000 species. Hardly any of these had been seen before. Along with viruses were signs of huge numbers of species of bacteria and other organisms.

The approach has also been taken up by Craig Venter (1946–), a biologist and pharmacologist who founded The Institute for Genomic Research (TIGR) in Rockville, Maryland, and later the company Celera Genomics. (Both organizations played an important role in the Human Genome Project, completing a second version of the genome at the same time as the international public version.) Since leaving Celera at the completion of the project, Venter has been sailing the world in a 95-foot (29-m) yacht called *Sorcerer II,* sampling the world's oceans and investigating other environments—including the human body. In a lecture given in 2007, Venter summed up the dis-

Craig Venter, founder of The Institute for Genomic Research and Celera Genomics, has dedicated his recent efforts to metagenomic studies of the world's oceans. *(Craig Venter and Liza Gross)*

coveries by stating, "Earlier this year . . . we published a single scientific paper describing more than 6 million new genes. This one study more than doubled the number of genes known to the scientific community and the number is likely to double again in the next year."

Metagenomic studies also provide a deep look at the effects of natural selection. Different environments are unique. Each one is more advantageous to some types of biochemical processes than others. A 2002 study of farm soil carried out by bioinformaticist Peer Bork and his colleagues at EMBL revealed a wide range of genes involved in the way different species respond to each other; it also turned up dozens of molecules involved in breaking down plant material. None of these were found in whale bones, taken from the ocean floor, or samples from the Sargasso Sea. The latter region yielded hundreds of genes similar to bacteriorhodopsin, a pigment which responds to light and allows cells to snatch energy from the environment.

A typical metagenomics study reveals thousands of types of genes that have never been seen in laboratory species. Some represent biological processes that do not occur in human cells. Investigating what these molecules do in exotic species of microorganisms will keep scientists busy for years to come. Researchers expect that many will have applications in medicine and industry.

This type of research is necessary to establish a baseline by which to measure the effects of global warming, human activity, or other types of changes on the environment. Researchers know that the growth of human populations, deforestation, and chemical and biological pollutants have caused the extinction of a large number of species, and the rate at which this is happening is increasing. The International Union for Conservation of Nature (IUCN) regularly evaluates the risk of extinction faced by thousands of species. Of the 40,117 species that were being monitored in 2006, more than 40 percent were listed as threatened. Only metagenomic approaches can determine whether the same phenomenon is occurring at the level of microorganisms.

Metagenomics, Extremophiles, and the Search for Extraterrestrial Life

For years, the evidence had been accumulating. Finally, on June 20, 2008, NASA confirmed the discovery: The Phoenix Mars Lander had dug a small hole and exposed a bright patch just below the surface that turned out to be water. This raised hopes that the planet might also hold living organisms. Many researchers believe that if water is present, there is a good chance that life will evolve, and in the past the surface of Mars held enormous quantities of water. An array of instruments would be used to analyze a small sample of soil, looking for signs of organic activity. But extraterrestrial microbes would undoubtedly have a completely different biology than Earth organisms. If alien life existed on Mars, would machines built by humans detect it? Not knowing exactly what to look for, scientists have been using the results of metagenomics studies and research into organisms called extremophiles that live in environments other forms of life are unable to cope with in hopes of obtaining hints about alien biology.

A major difficulty in searching for life beyond the Earth is that every living organism on this planet—from the simplest bacterium to human beings—shares a core set of molecules and biological processes. For example, all species store hereditary information in the form of DNA, which they transcribe into RNA molecules, which are then used as templates to make proteins. Cells have to be able to copy their DNA, convert the raw materials in food into useable molecules, and respond to changes in the environment. The universality of these processes on Earth makes it hard to imagine a form of life based on another chemical system.

An alien biology might not be totally different, however. Some researchers believe that extraterrestrial organ-

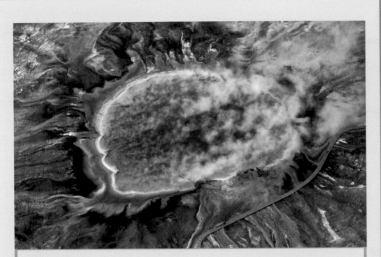

Metagenomic studies of extreme environments like hot springs and the organisms that live there may show scientists what to look for in the search for extraterrestrial life. This is the Grand Prismatic Spring, a hot spring in Yellowstone National Park, which is home to archaeal cells. *(Jim Peaco, National Park Service)*

isms might share some common elements with life-forms on this planet. Experiments attempting to reproduce the environment of the very early Earth, before life arose, have shown that some of the building blocks of proteins can arise from inorganic conditions. Even when the conditions have been changed, amino acids are almost always detected; they have even been found in clouds of gas in deep space. Amino acids are not alive, and the laboratory experiments have not produced entire proteins, DNA, or RNA, which are much more complex. That is not surprising, because it may have taken hundreds of millions of years for these molecules to arise in the vast laboratory of the early Earth's oceans. On the other hand,

(continues)

(continued)

the production of the first RNA and DNA molecules may have been a unique event that happened only on Earth. (It might also have taken place in space, and the molecules arrived on Earth as hitchhikers on meteorites, seeding the planet.)

It is also possible that extraterrestrial life might be based on an entirely different chemistry, particularly in environments that are quite different from the Earth. The surface of Mars, for example, is composed of 14 percent iron—nearly three times the amount in the Earth's crust. Organisms here make use of iron, taking advantage of its high chemical activity; at the same time, they have to control it carefully because even a slight overdose will disrupt cell chemistry. Life on Mars would need to have mechanisms to cope with this.

It might be possible to get a glimpse of what it would be like by studying organisms that live in iron-rich environments on Earth. Extremophiles such as bacteria or archaea that live in the boiling waters of hot springs or soil with high amounts of salt need special mechanisms to cope with such conditions. Extraterrestrial organisms may have adopted similar chemical strategies, which would give NASA an idea of what to look for.

Before they can hope to understand alien biology, researchers may need to get a better grasp of what is hap-

Scientists universally agree that a drop in biodiversity could have extremely serious effects on humanity. Over half the pharmaceutical compounds that have been developed for use in the United States are derived from compounds found in plants, animals, and microbes. Insects—a huge number of which are also threatened—play a crucial role in pollinating plants. It would be impossible to replicate that activity artificially.

pening on Earth. Metagenomic studies are revealing a range of processes that have never been observed in the laboratory because so few organisms can survive there. They are also revealing chemical signatures of different types of environments such as the farm soil and ocean floor. Living beings on a moon of Jupiter may have started with a much different chemistry than the cells born in the oceans of the early Earth, but natural selection may have pushed them to adapt in ways similar to organisms living in comparable environments here.

Until the first extraterrestrial is found, there is no way to really guess what it will be like, so the instruments needed to detect it are being designed in a very flexible way. The Phoenix Mars Lander is equipped with an instrument called a thermal and evolved gas analyzer (TEGA), constructed by the University of Arizona and the University of Texas at Dallas. The TEGA contains eight tiny ovens that will slowly cook small samples of soil, looking for changes in energy and chemistry. In the end, the temperature is so hot that the material turns into a gas, which is analyzed in a mass spectrometer. The hope is that the chemical experiments will reveal some sort of respiratory process—a conversion of substances that could only be carried out by a living organism—and that the mass spectrometer will reveal complex organic molecules. The latter may be detectable even if life vanished from Mars long ago.

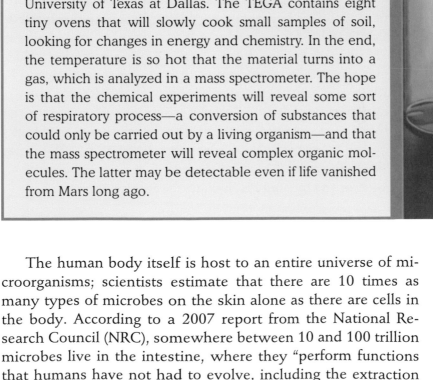

The human body itself is host to an entire universe of microorganisms; scientists estimate that there are 10 times as many types of microbes on the skin alone as there are cells in the body. According to a 2007 report from the National Research Council (NRC), somewhere between 10 and 100 trillion microbes live in the intestine, where they "perform functions that humans have not had to evolve, including the extraction of calories from otherwise indigestible components of our diet and the synthesis of essential vitamins and amino acids. The complex communities of microbes that dwell in the human gut shape key aspects of postnatal life, such as the development of the immune system, and influence important aspects of adult physiology, including energy balance. Gut microbes serve their host by functioning as a key interface with the environment; for example, they defend us from encroachment by pathogens that cause infectious diarrhea, and they detoxify potentially harmful chemicals that we ingest."

MANIPULATING GENES

The tools of genetic engineering introduced in chapter 1 have given rise to a wide palette of methods by which researchers can manipulate genes. The previous chapters have described some of the most important applications: creating crops resistant to pests or herbicides and using bacteria or animals to produce proteins for medical use. Yeast cells have been altered to yield better beer, and researchers are using bacteria to clean up environmental contamination through bioremediation. The late 1990s and early 21st century have seen the development of several new methods to give researchers a much more precise control of genes. At the moment, these techniques are mainly being used to discover gene functions, but one day they may produce new types of medical therapies.

Some of the techniques include the following:

- knock outs, which delete a gene
- *knock ins,* which add a gene to a cell or organism that does not normally have it
- overexpression studies that raise the amount of RNA and/or proteins produced from a given gene

Hermann Muller was the first researcher to deliberately introduce mutations in animal genes, using X-rays that caused random changes in DNA bases. Later, scientists began to use chemical mutagens. These measures often lead to knock outs by creating a gene sequence that encodes a defective protein. Another effect may be *constitutional activation,* in which a gene is switched on even when it should not be. If the molecule is involved in stimulating cell division, it should normally be turned off; an overactive version may lead to a tumor.

Modern reverse genetic techniques allow scientists to make precise, targeted changes in specific genes and watch what happens to cells, plants, or animals. The first knockout methods were all or nothing, completely removing a gene and eliminating its functions in all of an organism's cells throughout its lifetime. If the molecule plays an important role in an animal's embryonic development, this likely leads to such severe defects that the fetus dies before birth. Obviously, that makes it impossible to study a gene's functions during later phases of its life. It is often the case that the same gene may be needed at different times to do different things in various types of cells. For example, a protein called PS1 seems to act as a switch for different types of functions: It is needed to pass important signals that tell some types of cells to grow and develop. At other times and places in the body it is involved in *apoptosis,* a type of cell suicide that is necessary as tissues form. If the gene for PSI is removed, organisms lose control of these processes.

It is not surprising that proteins have multiple functions or even tasks that may seem contradictory. Human genes evolved from much smaller genomes in ancient ancestors with much simpler bodies and fewer genes, and those ancestors can all be traced back to a single cell. The development of new cell types and tissues did not necessarily require that new genes arise; often it occurred because cells began using the existing set of molecules in new ways. Just as a variety of electronic devices have some of the same components, multiple systems in the body rely on common proteins that have adapted to different tasks. So yeast, which is a single cell, contains proteins that now help build brains, eyes, and other highly complex organs in animals.

Getting a handle on fundamentally important genes required a way to shut down genes in specific types of cells at specific times. In the mid-1990s, the geneticists Klaus Rajewsky, Frieder

Schwenk, and their colleagues at the University of Cologne in Germany figured out how to do this with the invention of *conditional mutagenesis*. Their method relies on the fact that genes are accompanied by control elements—sequences that proteins dock onto in order to control when they are activated. Rajewsky's lab built genes with artificial switches that gave the scientists control over when and where a gene was shut down in an organism.

As with many other methods in genetic engineering, the technique is based on molecules from bacteria. An enzyme in bacteria called Cre recognizes patterns in DNA called loxP sequences. If Cre finds two of these sequences in DNA, it binds to the sites and draws them together, making a loop of the DNA that lies between them. This looped sequence is cut out, destroying a gene or any other information that it contains (such as regions that control a nearby gene). The cell then repairs the break by gluing the cut ends together. Thus, the first step in creating a conditional mutant is to build an artificial gene centered between loxP sequences.

The DNA is only knocked out in cells that produce both Cre and loxP sequences. If they are active in all of an organism's cells, the effect is like a complete knock out. Since the whole idea behind conditional mutagenesis is to avoid this, Rajewsky and his colleagues had to find a way to activate Cre only in particular types of cells. The solution was to find other genes that were only switched on in specific tissues or cell types, such as a molecule which is only produced in the brain. By combining Cre with the control regions of such genes, scientists could ensure that it too would only become active in the brain. The same technique could be used to study genes in any other tissue, providing a unique control region could be found.

Further refinements now allow scientists to determine the time as well as the type of tissue in which Cre becomes active. This is accomplished by attaching a switch, such as a receptor protein called LBD, to the Cre gene. LBD only becomes active in the presence of a specific hormone. Since animals do not naturally produce this hormone or obtain it through their diets, Cre

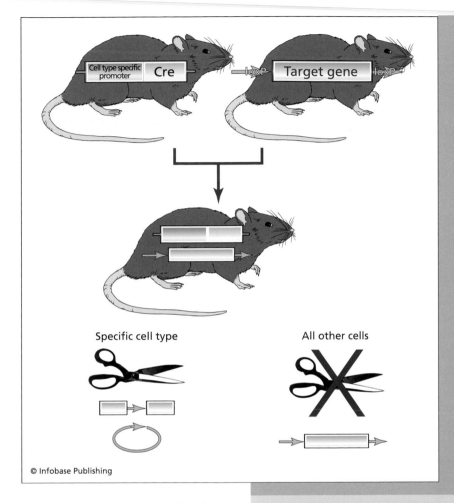

Cell type specific promoter | Cre

loxP | Target gene | loxP

Specific cell type

All other cells

© Infobase Publishing

Conditional knock outs using Cre. This method requires developing one strain of mouse with the Cre gene—which acts like a pair of scissors—and a second strain with the target gene (blue). When the two mice are mated, they produce a mouse with the scissors and the gene that it will remove. By placing the gene for Cre behind a promoter that activates it only in a certain kind of cell (green), scientists can knock out the target gene only in that type. The rest of the animal's cells will keep the target gene.

remains inactive until the desired time. Then the hormone is administered in an animal's food or by injection.

The Cre and loxP marked genes have to be introduced into separate strains of mice, which are then mated. Some of their offspring will have cells with both Cre and the targets. While this means waiting at least two generations for a mouse that has both elements, it also permits scientists to mix and match strains with Cre under the control of different tissues with those with different genes marked by loxP. For example, with the same Cre mouse, researchers can investigate the functions of different molecules in the brain by mating it with animals that have loxP attached to different target genes. And the reverse is also true. If the same protein is needed in the brain and the kidney, for example, and its gene has been tagged with loxP, scientists can mate the mouse with one Cre animal to test its functions in the brain and another to see what it does in the kidney.

Ideally, researchers would like to have a strain of mouse in which each gene is surrounded by loxP elements and other strains that express Cre in each tissue and cell type. Theoretically, this would allow researchers to test the function of every gene in every kind of tissue. It would be an enormous amount of work—mice have at least 13,000 genes and at least several hundred different cell types. Yet the usefulness of the mouse in creating human disease models has convinced many researchers that doing so could be worth the effort. This has encouraged scientists to start creating Cre zoos—collections of animals expressing Cre in different tissues. These animals are commonly shared by different labs, saving time and reducing the number of animals used in research. Centralized collections of mouse strains have been established at Jackson Laboratories in Maine, the European Mutant Mouse Archive near Rome, Italy, and elsewhere.

Studying these animals will not solve all questions about the functions of genes in humans or even in mice, because everything that happens in cells and organisms requires the collaborative efforts of many genes. One day, scientists dream of developing research animals with switches on every gene; this would give them control of many genes at the same time and allow them to investigate complex patterns.

In the meantime, researchers have discovered other ways to shut down genes—for example, by introducing small RNA molecules that block the production of proteins from other RNA molecules (see section titled Genomic Technologies on page 91). If researchers can find a way to introduce these molecules

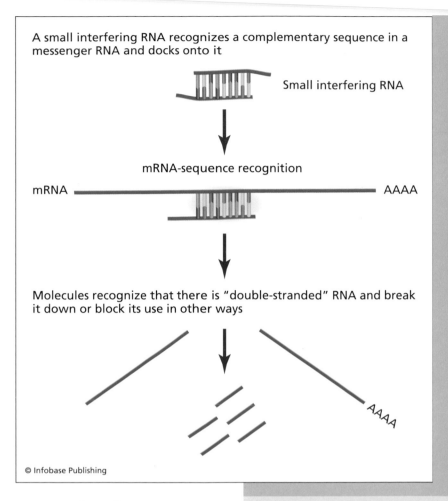

A small interfering RNA recognizes a complementary sequence in a messenger RNA and docks onto it

Small interfering RNA

mRNA-sequence recognition

mRNA ———————————————————————— AAAA

Molecules recognize that there is "double-stranded" RNA and break it down or block its use in other ways

AAAA

© Infobase Publishing

into specific cells in an organism, it would give them a useful new knockout technology that might even be turned into medical therapies.

The first successful clinical trial of such *small interfering RNA* (siRNA) molecules was carried out in 2005 by Sirna Therapeutics, a pharmaceutical company based in Boulder, Colorado, on patients suffering from macular degeneration. People with this disease lose their eyesight because of the death of vision cells called rods and cones or

Small interfering RNAs (siRNAs) provide a new method of knocking out genes. When siRNA are introduced into cells, they dock onto a messenger RNA with a complementary sequence and prevent it from being used to make the protein it encodes.

because blood and proteins leak into the inner lining of the eye. Doctors treated the patients with siRNAs to block the production of a protein that plays a key role in the disease. The experiment was considered safe because the eye is relatively closed off from surrounding tissues, which meant that there was little danger that the molecules would escape to other parts of the body. All of the patients showed improvements over the course of 157 days with no signs of side effects. More clinical trials are planned to treat other diseases.

IMAGING MOLECULES WITHIN LIVING CELLS

Google Earth allows an Internet user to zoom in on any region of the world, close enough to pick out houses, cars, and people. But the resolution is not high enough to identify a single person. Similarly, the most powerful electron microscopes can sometimes spot large single molecules and complexes containing many of them, but the image is not sharp enough to directly identify a specific protein. That would be helpful in understanding the functions of single molecules and the processes that go on in cells.

The resolution of light microscopes is about 1,000 times poorer than that of an electron microscope. Even so, light microscopy has been undergoing a renaissance because of the discovery of fluorescent proteins that can be used in new ways to study living processes.

In the early 1960s, Osamu Shimomura (1928–) and his colleagues at Princeton University isolated two luminescent proteins from *Aequorea victoria,* a species of jellyfish. Doing so required that Shimomura undertake a seven-day drive from New Jersey to Friday Harbor Laboratories of the University of Washington to capture and then dissect 10,000 jellyfish in just a few months every summer. Shimomura and his colleagues managed with the help of schoolchildren. By the mid-1970s, the researchers had perfected their routine and were collecting more than 3,000 jellyfish every day. In 1986, laboratories in the

United States and Japan simultaneously isolated the DNA sequence encoding one of the proteins, called aequorin. The second, called *green fluorescent protein* (GFP), was isolated in 1992 by Douglas Prasher and colleagues at the Woods Hole Oceanographic Institution in Massachusetts. These were crucial steps on the way to being able to work with molecules and turn them into tools for research.

GFP absorbed blue light emitted by aequorin and flashed brilliant green. Martin Chalfie of Columbia University immediately realized that it might be possible to turn the protein into a tool for microscopy. A laser microscope could play the role of aequorin; shining the right wavelength of light on GFP might trigger it to flash. The real use of the tool would come from the fact that the light-emitting part of the protein was located in one small, compact region (or domain) of the molecule. It might be possible to attach the domain to other proteins. If so, it would serve as a beacon that would allow molecules to be tracked under the light microscope.

Several steps were necessary to turn GFP into such a tool. S. James Remington, a structural biologist at the University of Oregon, obtained crystals of GFP and used X-rays to obtain a high-resolution map of the molecule. One of Remington's collaborators, Roger Tsien, and his colleagues at the University of California, San Diego, discovered a way to alter the molecule's structure so that the wavelengths of light produced by laser microscopes could activate it. Additional changes made the module much brighter and allowed GFP to work efficiently at body temperature. Since the late 1990s, Tsien's laboratory and others have developed versions of the molecule that emit other colors, including blue, cyan, and yellow. And, using proteins from coral and other animals, more fluorescent tools have been produced, which can be attached to other genes that make proteins that are fluorescent but otherwise normal. For their work, Shimomura, Chalfie, and Tsien were awarded the 2008 Nobel Prize in chemistry.

The method has several important uses. The first is simply identifying whether a molecule is made by a particular type of cell and where it carries out its functions: in the nucleus, the membrane, or another region of the cell. Another use of fluorescent microscopy is to find out whether switching on one gene leads to the activation of another—by marking them with different colors and watching to see when they are made.

Some of the more elaborate uses of GFP and similar proteins are based on the physics of how they absorb energy. Each GFP-tagged protein gives off a particular signal with precise characteristics. The signal shifts whenever the protein's activity changes—for example, when it binds to another protein or a small substance such as a drug. Thus fluorescence microscopy has become an important part of the drug development process as well as a tool to investigate the functions of molecules. If two molecules are labeled with different fluorescent modules and they bind to each other, each absorbs a bit of the light energy given off by the other. This can be detected by measuring the light that they emit. For the first time, researchers could directly observe the binding of two proteins in a living cell—even though the molecules themselves are too small to be seen in the microscope.

Next, researchers learned to apply the methods to tissues and organisms. The light from a microscope's laser can penetrate several layers of cells and excite a fluorescent molecule that lies below the surface. This principle was used in the late 1980s in the development of confocal microscopes. These instruments use a point of laser light to scan a sample, focused on a particular depth. It excites the fluorescent proteins; then the laser is refocused on the next lower layer. An image is captured of each layer and then a computer assembles the slices into a three-dimensional image. As digital imaging technology improved and computers became faster, it was possible to do this with living samples, such as insect larvae or fish embryos suspended in a liquid. A favorite subject was the zebrafish, a tiny fish that was becoming popular for genetic experiments. The fish is virtually transparent, which makes it easy to observe internal structures over the course of development.

Before these methods can be used, a scientist has to use genetic engineering techniques to attach GFP or another fluorescent module to specific genes. This means that molecules

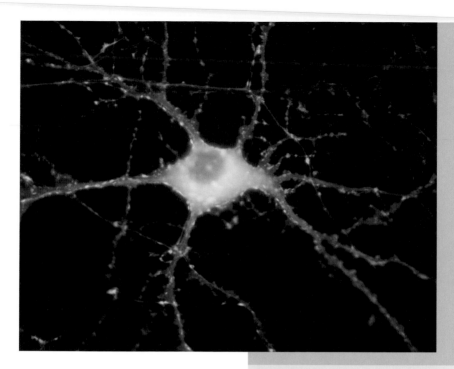

have to be investigated one by one; researchers cannot peer into a sample and survey all the proteins that are found there. A new method gets around this restriction by combining microscopy

Molecules in this neuron have been labeled with a fluorescent protein, allowing researchers to observe their locations and behavior using light microscopes. *(Michael A. Colicos, Division of Physical Sciences, USCD)*

with mass spectrometry. A beam is passed through a thinly sliced sample, creating ions that are captured by the spectrometer. They are weighed, as described earlier in the chapter, and from the results it is possible to identify the proteins that are present in the tissue.

THE DIGITAL EMBRYO

The early 21st century has seen the development of a range of techniques that can be used to study the activity of molecules and other biological processes in living organisms. One of the most interesting new applications is to watch developmental processes in embryonic fish and other laboratory organisms under the microscope over long periods of time. An ongoing project at the European Molecular Biology Laboratory (EMBL) in Heidelberg, Germany, is doing this in a unique way.

In the 1990s, Ernst Stelzer and his laboratory at EMBL gained a reputation for developing innovative types of microscopes that have helped bring light microscopes into the molecular age. Alongside making improvements in confocal instruments, they have constructed new types of microscopes that can make high-resolution, three-dimensional films of living fish and other small embryos.

The recent project is a collaboration between the lab of Stelzer and the developmental biologist Jochen Wittbrodt, one of his colleagues at EMBL. Wittbrodt wondered whether it might be possible to carry the method even further and study a single organism over long periods of time, perhaps even following its entire embryonic development. The embryo could be kept alive in a small, water-filled chamber that served as a sample chamber. But first Philipp Keller and his colleagues in Stelzer's group needed to make some improvements in the microscope. One problem was blurring and shadows that prevented scientists from obtaining a sharp look at details below the surface of a sample.

The problems stemmed from the fact that the laser of the microscope and the detector that captured images were right next to each other, aimed at the sample from the same direction. This was like taking a picture of an aquarium using a digital camera with a flash mounted right next to the lens. The glass of the aquarium might pose a problem by reflecting the flash; in the same way, a fluorescent molecule at the top of a sample might interfere with seeing molecules that are underneath it. Additionally, the picture of the fish inside might be very sharp, but it would be hard to guess whether they are at the front of the aquarium or the back, because most cameras do not provide very good information about how far away something is. Fish at various distances might be in focus. Researchers were having the same problem with laser microscopes when they wanted to make three-dimensional images—the resolution from side to side was very good, but along the line of sight, things were much blurrier.

Photographers sometimes get around these problems by using a remote flash, aimed at the subject from another angle, and the solution developed by Keller and other members of Stelzer's group was similar. In their new method, called digital scanned laser light sheet microscopy (DSLM), the microscope lens examines a specimen from the front while light enters from the side. The light consists of a very thin sheet that is slowly passed through an organism, from front to back. Only objects within that sheet are illuminated, which tells the researcher exactly how far away they are. The sample is then rotated and illuminated from different directions. This produces sharp slices that can be assembled into three-dimensional images with the help of the computer.

Keller and Wittbrodt began using the method with embryos of zebrafish and another small fish called medaka. One aim was to make very detailed maps revealing the tissues that produced specific molecules at various stages of development. Then they embarked on a much more ambitious project to create a living genealogy of each of the cells needed to make up a fully formed fish.

This has been an aim of biology since the mid-19th century, when Rudolf Virchow proposed that every adult cell stems from a single, fertilized egg. The best way to understand development would be to be able to trace the complete life history of every adult cell, back through each stage of specialization and development, all the way back to the egg. Researchers had some basic techniques to do this. In the late 19th century, embryologists learned methods to stain particular types of cells in the early embryo. Each time such a cell divided, it passed along the stain to its offspring. But these methods were imprecise because they required dissecting the embryos; cells could not be tracked in a single, living organism. And even simple animals consist of so many cell types that following them all would be impossible. According to Wittbrodt, a comprehensive study of development in the small worm *C. elegans* requires tracking only 671 cells, whereas "the analysis of complex vertebrate species requires the simultaneous determination and tracking of the positions of tens of thousands of cells."

The limitations left open many questions about developmental processes. Wittbrodt hoped that some of them could be answered using the new microscope, which could observe the same embryo for several days. This posed significant technical and computational challenges. "In order to observe and follow the nuclei of, e.g., the 16,000 cells of an 18h-old zebrafish embryo," Wittbrodt writes, "a volume of $1000 \times 1000 \times 1000$ cubic micrometers must be recorded at least once every 90 seconds." This was the maximum amount of time that could be taken to make a complete scan of the embryo. If it could be done, it would give the researchers a smooth, rolling film that would allow them to track each of the cells in the embryo. Cells divide and migrate quickly in early embryos. They can move away from their original positions in 90 seconds, but they do not move far enough to get lost. At longer intervals, the computer would lose track of them.

Additionally, the researchers wanted more than just a series of images. They wanted to teach the computer to identify single cells and keep track of them, then record each position in a database. This would provide a digital representation of each stage of the embryo's development that could be used to create hypotheses about the influences of genes on cell behavior.

The light sheet had to be moved at tiny increments through the sample, resulting in about 400 image slices that had to be combined to create each frame of the three-dimensional film. For a 24-hour recording, this added up to 400,000 high-resolution images per embryo that had to be combined in the computer.

In 2008, the machine was ready, and the data was captured. The group began sifting through the data. Each experiment recorded the history of an embryo from the first egg to a point at which it consisted of about 20,000 cells. The project shed new insight into some key events in embryonic development, such as gastrulation. This is the process by which a ball of undiffer-

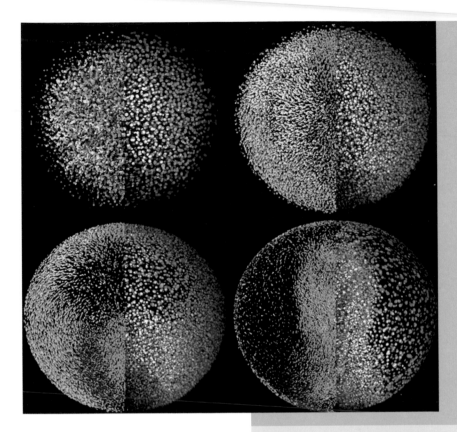

entiated cells forms three specialized layers, each of which
goes on to form specific tissues and organs. At a very
early stage, when the embryo
has taken on the shape of a
ball, made of a thin layer of
cells, some of them migrate
inward through a gap in the
surface to create new layers.
The study showed that this
migration happens through
a sudden influx of cells—like
grains of sand pouring through
a hole—rather than as a continuous, rolling-under movement of
a sheetlike layer of cells.

Using a new microscope to create a
digital embryo. These images were
made by Philipp Keller, a student in
the groups of Ernst Stelzer and Jochen
Wittbrodt at the European Molecular
Biology Laboratory in Germany.
The digital scanned laser light sheet
microscope, described in the text,
makes three-dimensional films of fish
embryos from the earliest stages of
development up to the formation of
clear body structures. Each cell can be
tracked in the computer. The right half
of each image is the actual microscope
picture; the left half is a computer
reconstruction. (Philipp Keller, EMBL)

Wittbrodt and his colleagues think that the method has the potential to answer a wide range of questions about animal development and disease. It will allow them to track any cell from its origins to its final form, when it is fully specialized and embedded in an adult tissue. The researchers can observe the formation of organs in real time. Some of the most interesting applications, Wittbrodt says, involve watching how development occurs in organisms with genetic defects. For example, proteins on the surfaces of neurons direct the way the cells form networks in the brain. Very subtle changes alter these connections and can cause the brain to develop in a different way. This process of hardwiring can be observed as it takes place in the embryo using the new microscope, and it may provide insights into the causes of some types of brain damage that have been linked to genes.

IMAGING BODIES

Large animals or human beings cannot be put under a microscope for observations; nor are they transparent or small enough to be studied using the DSLM instrument described in the previous section. But a number of other new techniques are available to capture images of the behavior of living cells and tissues in bodies. These methods are currently being used in medical applications such as visualizing tumors without surgery, discovering how much damage has been caused by a stroke, or simply exploring the function of the brain.

The earliest noninvasive technology that could be used in patients was X-rays. The German physicist Wilhelm Röntgen (1845–1923) discovered that when these waves were shined through a body, they were blocked or absorbed by various types of tissues in different ways. Capturing the absorption pattern on a photographic plate, behind the body, yielded an image in which bone, fat, and muscle could be clearly distinguished. X-rays gave doctors a look at broken bones and problems in other tissues, particularly the lungs.

Using X-rays in conjunction with substances that absorb them, such as barium, gave doctors a closer look at soft tissues through which the substances move. These images were two-dimensional up to the development of *computerized tomography* (CT) scans, initially based on a moveable X-ray machine attached to a scanner. The method was simultaneously announced in 1972 by the British researcher Sir Godfrey Hounsfield (1919–2004) at the Central Research Laboratories of Electronic & Musical Industries Ltd. (EMI) and Allan Cormack (1924–98) of Tufts University in Massachusetts, leading to the joint award of the 1979 Nobel Prize in physiology or medicine. (Best known for its record label, which featured albums by the Beatles, EMI was deeply involved in technological research and had helped build radar devices during World War II.)

The instrument is passed over the body, creating image slices that are assembled by a computer into a three-dimensional image. The method is limited because barium or hard parts of the sample block the passage of the waves, preventing researchers from seeing the soft tissues in much detail. The original CT machines relied on X-rays. Even though small doses were used, there was always a small risk of damage to patient tissues.

In the meantime, CT was being used with other types of imaging. *Magnetic resonance imaging* (MRI) was also the product of applying discoveries in physics to a biology problem. The principles behind the technology were discovered in the 1930s by Isidor Rabi (1898–1988), a physicist at Columbia University in New York City. He was using molecular beams to investigate the forces that hold electrons to the nuclei of atoms, work which earned him the 1944 Nobel Prize in physics. It also had practical applications in the development of radar, a project that Rabi was recruited to work on in World War II.

Two physicists who had also been recruited, Felix Bloch (1905–83) and Edward Purcell (1912–97), adapted nuclear magnetic resonance (NMR) so that it could be used to investigate liquids and solid objects, an accomplishment recognized with the award of the 1952 Nobel prize in physics. NMR slowly became an important method for the determination of structures of biological molecules thanks to the efforts of Kurt Wüthrich (1938–), a Swiss chemist who now heads laboratories at the Swiss Federal Institute of Technology in Zürich and the Scripps

Institute in La Jolla, California. Wüthrich began working with NMR when he joined Bell Laboratories in Murray Hill, New Jersey, in 1967, where he had access to one of the best instruments in existence at that time. He used it to study the structure and behavior of proteins in liquids; thanks largely to his efforts, it has become a standard tool in structural biology, drug discovery, and related fields. The 2002 Nobel Prize in chemistry was awarded to Wüthrich for this work.

Unlike X-rays or other imaging methods, MRI does not require radiation or ionization that might be harmful to living cells. Theoretically, patients can be scanned again and again with no adverse health effects; the only risk is that tissues are slightly heated and the procedure should not be used on people fitted with pacemakers or other electronic devices.

The method is based on exposing a sample or patient to strong magnetic fields. The nuclei of atoms are sensitive to these fields and absorb the energy. This effect is like placing a strong magnet next to a compass and then removing it. At close proximities, the magnetic field draws the needle of the compass. Removing the magnet again makes the needle return to its normal position. Something similar happens with atoms in an NMR experiment. The magnetic field aligns their nuclei. When the field is relaxed, they snap back to their normal state, but the way that they do so depends on what other atoms are nearby. Applied to a tissue sample or an organism, this produces an image that is particularly good at showing differences in soft tissues or liquids.

The method can also be tuned to detect the presence of particular substances or molecules. One use has been to follow the activity of the brain as it performs different tasks. MRI shows changes of blood flow to various regions, which has been associated with brain functions.

MRI has also been adapted in clever ways to other types of problems. A recent study by the laboratory of Peter Schlag, a German cancer researcher at the Charité University Medical School in Berlin, proposes a use that might help doctors diagnose the severity of breast cancer. Traditional mammography depends on X-rays to reveal abnormalities in breast tissue. By showing the size, location, and shape of a tumor, mammographies help doctors look for signs that a tumor is undergoing transformations that lead to metastases. But the method is imprecise, Schlag says, and it cannot be used to look for molecular events that might allow a doctor to predict how the tumor will develop. Magnetic resonance mammography, on the other hand, might provide more information.

Schlag's new method is based on the fact that molecules that can be detected and tracked by MRI might behave differently in tumors and noncancerous tissue. For example, rapidly growing tumors stimulate the formation of new blood vessels; tumor cells are just as dependent on nutrients delivered through the blood as healthy cells. But they grow much more quickly than other adult tissues, which means that special mechanisms are required to meet their demands. The adult body often builds new blood vessels—to repair damage caused by injuries, for example—but this happens at a slow rate, and the process is carefully regulated.

Tumors have to overcome the regulatory systems, and in doing so, they create vessels that are slightly different than those of surrounding healthy tissue. "One of these differences," Schlag said in a personal interview with the author, "is that the new blood vessels are not sealed as tightly as other vessels. Gaps between the cells permit large molecules to slip through and seep into the surrounding tissue. This led us to start looking for a substance of the right size—able to slip through tumor-related vessels but not healthy ones—that can be detected by MRI. We found such a substance in indocyanine green, a fluorescent molecule that leaks into tumors but not into healthy tissue. Most of the substance stays in the bloodstream and moves to the liver, where it is cleared from the body. After this happens, the tumor has absorbed the molecule and stands out in high contrast against the surrounding tissue. In the patients we examined, the contrast is highest in malignant tumors, those which have metastasized or will do so soon."

The same approach can show how the body absorbs other substances, such as drugs. MRI is being used to look for substances that can cross the blood-brain barrier, a defensive mechanism that has evolved in the brain to protect it from toxins, viruses, and most parasites. The blood vessels of the brain are especially tightly sealed. This means that many drugs that are successfully delivered through the bloodstream to other parts of the body do not reach brain tissue. MRI can be tuned to pick up the traces of particular substances. By tracking their presence in blood vessels and surrounding tissues, the method can be used to find new substances that penetrate the blood-brain barrier.

4

The Future of Humanity and the World

"The advance of genetic engineering makes it quite conceivable that we will begin to design our own evolutionary progress," wrote the science fiction author Isaac Asimov in a collection of essays called *The Beginning and the End.* Even though genetic engineering had barely begun when the book appeared in 1977, Asimov and others clearly understood its potential. The idea that humans will take control of their own genetic future has been around ever since, and it has aroused both curiosity and fear. Part of the fear stems from the fact that researchers do not yet fully understand the human body and mind; some believe they never will.

Scientists have been manipulating human genes since the early days of genetic engineering—transplanting them to bacteria or other species, which can be used as factories for medically important molecules such as insulin—or manipulating lines of human cells grown in the laboratory. There is great interest in learning to replace defective genes in humans, as a cure for genetic diseases, but a safe and reliable method of doing so in embryos or adults has not yet been perfected. The idea of correcting the defects in egg or sperm cells or very early embryos has been rejected for ethical and technical reasons. Within a few years, the technical issues may be resolved, and it would likely be possible to extend the techniques commonly used in animals to humans—if society were to allow it to be done. One theme of this chapter is to explore the possible consequences of taking direct control of human genes and human evolution.

People are already shaping their future evolution. The driving force that changes species is natural selection, which implies that by altering the environment, humans will indirectly change themselves. The human impact on the environment involves technological inventions, changes in diet, overpopulation, genetically modified crops, pollution that causes global warming, and behavior that reduces the world's biodiversity. Over the long term, these factors will inevitably transform the species. This chapter looks at these issues from the perspective of some of the most exciting frontiers of genetics and biology. Each of these fields is so complex that it could easily fill a book of its own. Here, each topic will be introduced through the eyes of one of the most prominent thinkers and scientists working in the field.

DESIGNER BABIES AND A POST-HUMAN FUTURE

Until the publication of Darwin *On the Origin of Species* in 1859, nearly everyone regarded humans as far superior to any other living creature, the pinnacle of nature, nearly godlike. People expected to change in technological and social ways, and they hoped to improve themselves morally, but there was no real notion that humans might one day become an entirely different species. With the discovery of evolution and the laws of heredity, mankind suddenly saw itself as one small step along a branching, open-ended pathway of life, rather than an end point. The future of the species was suddenly up in the air. Evolution could not be stopped, so humans would inevitably change. Their descendants—if they survived—would likely be as different from the people of today as *Homo sapiens* is from the early primates that wandered the African savanna millions of years ago.

There is no way to predict what direction these changes will take as long as reproduction is left to the roulette of nature,

through which each new child is a chance mixture of parental alleles and a few new mutations. There is no guarantee that the people of the future will be smarter or more peaceful or that they will live in a healthier relationship to the environment. The only way to change this situation and steer evolution in a desirable direction may be through genetic engineering. But even if these methods could be used safely in humans, many ethicists and thinkers are concerned about the end result. Just as the eugenics movements of the early 20th century would not have worked—because they were based on a misunderstanding of human heredity and the nature of genes—deliberately manipulating the human genome might have entirely unwanted and unpredictable consequences.

In a 2002 book entitled *Our Posthuman Future: Consequences of the Biotechnology Revolution,* the American philosopher and political economist Francis Fukuyama (1952–) eloquently gives voice to these concerns. A decade earlier, Fukuyama gained widespread attention with his book *The End of History and the Last Man.* The title is not meant to imply that the species is doomed; instead, it raises a subtle question about whether people will need to write histories in the future. In the past, Fukuyama says, the historian's main focus was the struggle between competing political systems and ideologies. Such conflicts might soon be a thing of the past. Recent decades have shown that communism and most other types of regimes are unsustainable, he believes. With the arrival of Western liberal democracy, mankind has reached the logical end point of social evolution. "What we may be witnessing is not just the end of the cold war, or the passing of a particular period of postwar history," he wrote, "but the end of history as such: that is, the end point of mankind's ideological evolution and the universalization of Western liberal democracy as the final form of human government."

Just 10 years later, however, Fukuyama felt the need to refine his hypothesis. Discoveries in science—particularly biology and genetics—might change society or even human nature, which could well create the need for new forms of social organization. In *Our Posthuman Future,* Fukuyama points out that political systems are ultimately dependent on human nature, science, and technology. Liberal democracy is a product of human lifestyles and the mind—so what would happen if, for example, gene technology were used to change the brain?

These questions have led Fukuyama to consider the potentially dangerous and ethically questionable consequences of genetic engineering. Current uses of the technology include diagnosing serious birth defects, looking for genetic markers associated with diseases, creating genetically modified crops and animals, and developing therapies for health conditions by trying to repair or replace defective molecules. All of these methods are in their infancy, and most biomedical researchers believe that in the near future, they will be major tools in fighting diseases, such as cancer, and possibly in the repair of developmental defects caused by flaws in genes.

But the potential goes much further. If molecular medicine develops in the way most researchers expect, scientists will soon find genes linked to mental conditions such as schizophrenia and bipolar disorder. Research has also revealed connections between genes and intelligence, tendencies toward addiction, and other behavior such as attention deficit/hyperactivity disorder (ADHD). This condition is a good example of the concerns that Fukuyama and many others have about the potential uses of gene therapy—not because of ADHD itself, but because of the way the disease has been handled by the medical community and the family members of those affected.

The causes of ADHD are largely unknown. Symptoms include inattentiveness, hyperactivity, and a wide range of other behaviors that make it challenging for children to adapt to school and other social settings. Studies of twins reveal that there is likely to be a genetic component, although specific genes have not yet been identified. This means that there is no objective test to establish whether a person has the disorder. Diagnosis is difficult because it is usually hard to tell the difference between normal—although somewhat exaggerated—childlike behavior and the symptoms of ADHD. Studies of the rates at which the condition occurs in schoolchildren give widely varying results—from 2 to 14 percent. There is a large ongoing debate in the medical community about the disorder, motivated by concerns about misdiagnosis, ethical concerns about medicating children, and worries about the long-term effects of the drugs they are given. In spite of these issues, a growing number of parents are turning to medications to treat children that have been diagnosed with ADHD. While these treatments have certainly helped many children and their families, researchers worry that they might be overused. The way the drugs affect the brain is not completely understood, and there has been a lack of research into effects that they may have over the long term.

Fukuyama's concerns with genetics center around this issue of defining what is normal. He casts his mind toward the future and predicts a time when "Knowledge of genomics permits pharmaceutical companies to tailor drugs very specifically to the genetic profiles of individual patients and greatly minimize unintended side effects. Stolid people can become vivacious; introspective ones extroverted; you can adopt one personality on Wednesday and another on the weekend. There is no longer any excuse for anyone to be depressed or unhappy; even normally happy people can make themselves happier without worries of addiction, hangovers, or long-term brain damage."

This scenario, which sounds strikingly similar to the pleasure-intoxicated society of Aldous Huxley's *Brave New World,* is one of several ways that Fukuyama sees people adapting to and using the new possibilities of genetic science. Other medical discoveries, originally intended only as last-resort measures to save lives, have now been adapted for purposes like cosmetic surgery. If people have the chance to eliminate disease or improve themselves—making themselves more attractive or intelligent without serious side effects—what would keep them from doing so?

Giving parents control over their children's genes leads to other concerns; Fukuyama uses homosexuality as an example. One intensely debated topic has been whether sexual preference has a genetic basis or is determined almost entirely by environmental influences. While some studies have linked male homosexuality to a position on the X chromosome and others have reported differences between the brain structures of homosexual and heterosexual males, there is not yet a definitive scientific answer; scientists have not yet found a gay gene.

But suppose that such a molecule is found or that at least researchers discover a biological basis for homosexuality. Knowledge about the causes of things, Fukuyama writes, "will inevitably lead to a technological search for ways to manipulate that causality." In the case of homosexuality, he proposes a thought experiment: "Assume that in 20 years we come to understand the genetics of homosexuality well and devise a way for parents to sharply reduce the likelihood that they will give birth to a gay child. This does not have to presuppose the existence of genetic engineering; it could simply be a pill that provided sufficient levels of testosterone in utero to masculinize the brain of the developing fetus. Suppose the treatment is cheap, effective, produces no significant side effects, and can be prescribed in the privacy of the obstetrician's office." Even if homosexuality is completely accepted by society at this future date, Fukuyama wonders how many pregnant women would take the pill. He thinks that many would, even if they were not prejudiced against homosexuals. "They may perceive gayness to be something akin to baldness or shortness—not morally blameworthy, but nonetheless a less-than-optimal condition that, all other things being equal, one would rather have one's children avoid. . . . Wouldn't this form of private eugenics make them more distinctive and greater targets for discrimination than they were before? . . . Should we be indifferent to the fact that these eugenic choices are being made, so long as they are made by parents rather than coercive states?"

At the moment, parents cannot pick the characteristics of their children, except in the sense of prenatal screening for disease genes—a practice that is also troubling to many ethicists. The more choices parents have, the more expectations they are likely to have—that their children will be better behaved, will not become bald, or will develop in certain ways. However, human genetics is so complex that even if scientists decided to take drastic measures to alter the human genome, the children produced by genetic engineering would be unpredictable.

At the moment, people know (or quickly learn) that creating a new child is like spinning a roulette wheel: What happens is a bit of heredity, but mostly chance. The same thing will be true of engineered humans, but people may have a harder time adjusting when their expectations are dashed.

This modern form of eugenics—steered by genetic engineering—is only one of many ways that Fukuyama can imagine biotechnology influencing the future of human beings. "While it is legitimate to worry about unintended consequences and unforeseen costs, the deepest fear that people express about technology is . . . that, in the end, biotechnology will cause us in some way to lose our humanity—that is, some essential quality that has always underpinned our sense of who we are and where we are going, despite all of the evident changes that have taken place in the human condition through the course of history. Worse yet, we might make this change without recognizing that we had lost something of great value. We might thus emerge on the other side of a great divide between human and posthuman history and not even see that the watershed had been breached because we lost sight of what that essence was."

CURING BRAIN DISEASES AND THE NEW PHARMACOLOGY

At the beginning of the 21st century, the study of diseases of the mind and brain is entering a new phase. These diseases serve as a good example of the way medicine is being transformed by genetics and molecular biology. Researchers are starting to uncover the genetic and physical causes of numerous mental health problems, and this has significantly affected the way that the diseases and their victims are seen by the medical community and society as a whole. This is an enormous change from the state of things just a few decades ago, and the new perspective brought by the molecular revolution in the neurosciences will likely be the key to finding cures for Alzheimer's disease, strokes, and many other problems of the brain.

How far things have come can be seen in the fact that people with mental disabilities, particularly the residents of mental institutions, were among the most unfortunate victims of the eugenics programs of the early 20th century. They were often sterilized without their consent or even that of a relative. Under the Nazi regime, many were simply murdered. Systematic sterilization was finally halted in the United States in the 1930s because of ethical concerns and a recognition of patient rights. But at the same time, another equally questionable medical practice began making the rounds of hospitals and psychiatric clinics.

In 1935, the Portuguese physician Antônio Egas Moniz (1874–1955) began performing a type of brain surgery on human patients after learning that it successfully cured monkeys of aggressive behavior. Cutting a particular tissue in the front region of the brain—separating the white matter from the rest of the organ—rendered monkeys calm and friendly. Egas Moniz began performing the operation on humans. In many cases such *lobotomies* calmed the person, brought an end to epileptic seizures, or stopped other undesirable behaviors—because it broke connections that allowed impulses to spread from some regions of the brain to others. In the days before antipsychotic drugs such as Thorazine, the procedure was regarded as a huge breakthrough, and in 1949 Egas Moniz was awarded the Nobel Prize in physiology or medicine. In America, the neurologists Walter Freeman and James Watts streamlined the procedure so that it could be carried out routinely in psychiatric clinics, by pushing an icepick-like instrument into the brain through the patient's eye socket. Eventually, it was performed tens of thousands of times.

However, it often had terrible side effects. For example, the operation was performed on Rosemary Kennedy, the sister of President John F. Kennedy. The aim was to reduce what the family called aggressive behavior. Before the operation she was considered mildly mentally disabled; afterward, she was reduced to an infantile state. Other patients had similar problems or experienced undesirable personality changes. Long-term studies and a more careful examination of patients began to reveal many serious, undesirable side effects. Finally, the surgery fell out of favor as antipsychotic drugs became available.

Brain research and the way researchers searched for cures changed dramatically over the next decades, due to the arrival of noninvasive techniques such as magnetic resonance imaging (described in chapter 3). This shifted the focus of the work toward discovering how modules in the brain communicate with each other along electrical and chemical pathways to form thoughts, feelings, perception, and behavior. In disease, these conduits are interrupted. Imaging techniques have permitted the discovery of links between specific tissues and mental behavior and some ideas about how normal processes are disrupted in disease. Sometimes, this information can be woven into stories of how the brain functions. In her book *Mapping the Mind,* for example, the medical writer Rita Carver summarizes a current hypothesis of clinical depression: "Depression is caused by the firing of a circuit in which the amygdala feeds negative feelings to consciousness, the prefontal lobe pulls out long-term memories to match the feeling, the anterior cingulate cortex fastens on to them and prevents attention from shifting to anything more uplifting, and the thalamus keeps the whole circuit alive and firing."

Geneticists and molecular biologists, on the other hand, look at the brain from the bottom up, asking different sorts of questions. The contrast is most obvious in a case like Alzheimer's disease. More than 100 years ago, the German physician Alois Alzheimer (1864–1915) discovered that the brain of a woman who had suffered from dementia contained *amyloid plaques*: fragments of proteins that had formed tangled clumps in the space between brain cells. Eventually, the brain shrinks, and one-by-one its functions fail. Using imaging techniques, researchers can help diagnose the condition as it develops and watch the changes that take place.

Most of the work of molecular biologists has focused on the amyloid plaques, which block communication between the cells, prevent them from getting nutrients, and eventually cause their death. Only today are scientists beginning to understand how the plaques form. They begin as part of a protein called APP (amyloid precursor protein), which is made by neurons and some other types of cells. APP rests in the cell membrane, where it seems to play a role in memory and learning, but its healthy functions are not yet really understood. Quite a bit has been learned about its biochemistry, however, and some of the processes by which it contributes to disease.

Researchers have discovered that other proteins come along and compete to slice APP at various places, which leads to different types of fragments. Some of these are harmless, but others latch onto each other and form amyloid plaques that do not dissolve. Molecular biologists hope to learn why brains sometimes stop making healthy fragments and start producing the unhealthy form. The answer, they have learned, partly depends on which proteins do the cutting and the order in which they do so. Other molecules play a role by binding to the fragments and helping weave them into fibers.

It is possible that an existing drug—or one of the millions of compounds in the libraries of pharmaceutical companies—will block one of these processes, prevent the accumulation of the plaques, and stop the course of the disease. Knowing what to look for makes the search for treatments infinitely easier. Controlling the development of Alzheimer's disease will probably require learning to control the activity of these other molecules. That will be easier in some cases than others. Some families are particularly susceptible to the disease, probably because of the influence of other genes.

This is typical of the molecular age's approach to the study of other types of brain disorders and many other diseases. The first step in looking for a treatment is to understand how a problem affects cells and molecules. In a stroke, for example, the blood flow is cut off to cells in a particular region of the brain and they die. The reason for their death is not directly suffocation or starvation. Instead, the loss of the blood supply cuts off signals that cells need to survive and triggers other signals that tell them to die. In 2005, Oliver Hermann and Markus Schwaninger of the University of Heidelberg, Germany, discovered that in the wake of damage, a self-destruct signal is likely passed to cells' genes via a molecule called NF-kappa-B. Using mice that had suffered a strokelike condition, they showed that blocking the signal helps cells stay alive much longer and recover, even if the treatment comes a few hours after a stroke. Their finding

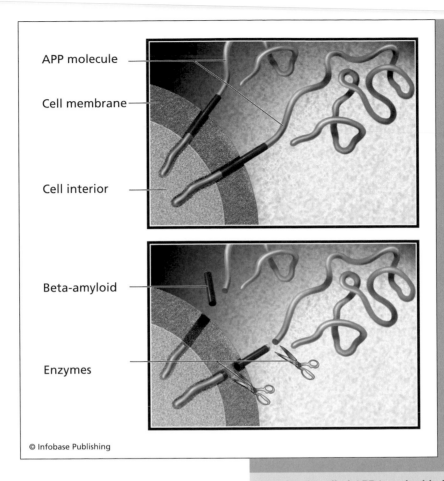

APP molecule

Cell membrane

Cell interior

Beta-amyloid

Enzymes

© Infobase Publishing

has been used as a starting point for the development of new, experimental therapies.

In the future, even the death of cells may not cause an irreversible health problem because there may be ways to replace them. Until the late 1990s, it was almost universally believed that nearly all of a person's neurons developed early in life; any that were damaged or died could not be replaced. In the meantime, researchers have discovered that the creation of new neurons

A molecule called APP is embedded in the surface of neurons, where it can be cut at various locations by enzymes. Some types of cuts (lower left) produce a fragment called Beta-amyloid that accumulates between nerve cells, causing the symptoms of Alzheimer's disease. Other ways of cutting the molecule (lower right) produce a harmless fragment.

continues into adolescence and later. In 1998, Peter Eriksson and other members of the laboratory of Fred Gage, working at Sahlgrenska University Hospital in Göteborg, Sweden, discovered that cells in a region of the brain called the hippocampus could differentiate into new neurons. Eriksson has gone on to show that additional areas of the brain hold stem cells that can do so. But with age the creation of neurons likely becomes rare, and it is also limited to specific parts of the nervous system. In a spinal cord injury, for example, cells are unable to repair breaks that prevent the brain from communicating with the rest of the body. But if biologists could tap into the potential of stem cells, by learning to control the signals that guide their specialization, it might be possible to activate the body's cells to do so.

In many other brain diseases, protein fragments accumulate and cannot dissolve. In Huntington's disease they collect in the cell nucleus rather than the space between cells—eventually with the same result, killing neurons. Under the right circumstances, so many different types of proteins form amyloid clusters that many researchers believe that eventually nearly anyone who lives long enough will suffer from one of these types of neurodegeneration. And the numbers are sure to rise if cures are found for other main killers that strike earlier, such as cancer and cardiovascular diseases. As James Thorson puts it in his book *Aging in a Changing Society,* "Remember, everyone has to die of something, so if one cause goes down, something else has to go up." Truly eliminating these diseases may require solving an even greater scientific question—why cells and bodies change over time—in other words, why people age. That theme is taken up later in this chapter.

Several new approaches are being taken to address the symptoms of neurodegenerative and other systemic diseases such as cancer. Regenerating tissues using stem cells is one. Additionally, of course, researchers continue to try to find natural substances or develop artificial compounds that can be used as drugs. The way that this is done has changed dramatically over the past 20 years and is increasingly moving toward what scientists call rational design.

This begins by pinpointing the cellular processes that cause a disease and identifying the molecules involved in it. The next step is to find a target: a particular protein or gene whose manipulation would give researchers a way to control the process. In Alzheimer's disease, for example, this might be one of the enzymes that cleaves the APP protein to generate a dangerous fragment. Once such a molecule has been identified, researchers try to obtain its precise three-dimensional structure (a picture of the arrangement of its atoms, which gives details of its shape and chemistry). This may reveal a location in the molecule for a drug to dock onto and change its activity. If that is successful, a researcher can scan databases of drug compounds in hopes of finding another molecule with the right configuration and chemistry, a molecule that can snap on, a bit like looking for the right electrical adaptor to fit a cell phone. Then the molecule is screened to see whether it really influences the activity of the target, and how well it does so. If an appropriate drug or compound is not found, researchers screen the target molecule against a library of substances and hope for a lucky hit. Promising candidates can then be rebuilt by chemists to be more effective.

Over the past decades, the methods to carry out such screens have been automated and improved, making it relatively easy for scientists to test thousands or even millions of substances for pharmacological activity. There are different types of screens. First, experiments are carried out in test tubes, looking for signs of changes in a molecule's chemical activity. Next, substances are introduced into cells, also using automated methods. If a molecule passes the test, it can be tried in animals; promising results may convince a pharmaceutical company to step in and conduct human clinical trials.

This is a much different strategy than scientists used in the past. It is based on an exact knowledge of molecular cell biology and the selection of a precise target, rather than simple trial and error and experiments that were often based entirely on animals. Researchers are continually on the lookout for new compounds that might make effective drugs. Other species—even unusual, exotic ones—are one important source; another is the

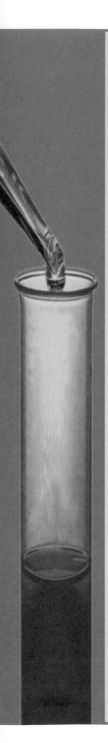

Cures from Shamans and Traditional Healers

Mark Plotkin (1955–   ), an ethnobotanist and president of the nonprofit organization Amazon Conservation Team, has dedicated his career to traveling to remote parts of the world, living with indigenous peoples and learning about the way their healers use plants. Plotkin's story is featured in the IMAX film *Amazon,* partly based on his 1994 book, *Tales of a Shaman's Apprentice.* His work is also the subject of a 2001 film called *The Shaman's Apprentice,* directed by Miranda Smith.

"Over thousands of years, through a method of trial and error, indigenous tribes have built up a storehouse of knowledge about the native vegetation," Plotkin writes. "There exists no shortage of wonder drugs waiting to be found in the rain forests, yet we in the industrialized world are woefully ignorant about the chemical—and, therefore, medicinal—potential of most tropical plants. . . . The approximately 120 plant-based prescription drugs on the market today are derived from only 95 species. A quarter of the prescription drugs sold in the United States have plant chemicals as active ingredients. About half of those drugs contain compounds from temperate plants, while the other half have chemicals from tropical species."

The Amazon alone is home to about 25 percent of the world's plant species. Why should organisms in such exotic locations be a source of interesting pharmaceutical compounds? One reason has to do with the equally varied species of insects that inhabit the rain forest. "The fact that the forest has not been devoured by this entomological onslaught is testament to these plants' abilities as chemical warriors. Plants protect themselves by producing an astonishing array of chemicals that are toxic to insects, thereby deterring predation. When ingested by

Amazon Rain Forest

VENEZUELA
COLOMBIA
GUYANA
SURINAME
FRENCH GUIANA
ATLANTIC
OCEAN
ECUADOR
PERU
BRAZIL
N
BOLIVIA
PACIFIC
OCEAN
CHILE
PARAGUAY
© Infobase Publishing
0   500 miles
0   500 km

Amazon rain forests are home to a rich variety of plant and animal species, many of which are found nowhere else on the globe. While these species potentially hold substances that can be used as medicines and for other purposes, the rain forests are shrinking at an alarming pace due to human activity, and many species are becoming extinct.

humans, these same plants—and their chemical weapons—may act in a variety of ways on the body: they may be nutritious, poisonous, or even hallucinogenic. And in some cases, they are therapeutic."

This is one reason that Plotkin and many other conservationists are racing against time to save the Amazon and other endangered ecospheres. There are many more (see the section titled Saving the World, later in this chapter). Tropical plants frequently contain alkaloids, natural chemical compounds that contain nitrogen and often have a powerful effect on the body. Caffeine, nicotine,

(continues)

(continued)

morphine, cocaine, and quinine (long the most effective treatment for malaria) are alkaloids.

Another natural substance that has currently attracted great interest among biologists is epigallocatechin-gallate (EGCG), found in green tea. The tea has been used as a remedy in traditional Chinese medicine for hundreds and probably thousands of years. Recently, Erich Wanker's group at the Max Delbrück Center for Molecular Medicine in Berlin, Germany, discovered that in the test tube, EGCG reduces the clustering of proteins responsible for Huntington's disease. "If EGCG was helpful in one case," Wanker told the author in an interview, "it might be helpful in others. So we have tested proteins responsible for Alzheimer's disease and Parkinson's disease. In each case, EGCG prevents protein fragments from clustering into fibers. We followed up in a second study to see if it would also have beneficial effects on cells. In each case, the result was positive." Wanker and his colleagues are now purifying and improving the substance, hoping that a form of EGCG can be developed which will slow or stop the progress of these brain diseases.

enormous range of plants or natural substances used by traditional healers.

Alongside stem cell therapies and the development of new pharmaceutical substances, the molecular age has spawned other new ideas about treating diseases. These methods include correcting defective molecules by delivering healthy versions to cells—in essence, giving the body new genetic instructions by which it produces its own therapeutic molecules. The biggest problem has been finding a way to get the molecules into cells, which have evolved defenses that protect them from taking up foreign genes or RNA. Some viruses manage to overcome these defenses, however, so one method that is being tried is *viral therapy.* This approach starts by taking a relatively harmless virus, removing any information that might cause an infection, and replacing its genetic material with the healthy form of a human gene. The virus is altered so that it cannot infect healthy tissues or be transmitted to another person. The technique has been tested in a number of clinical trials—in some cases very successfully, but there have also been deaths and negative side effects. Thus, researchers are searching for new ways to deliver therapeutic molecules to cells.

Another approach is to extract immune system cells from a patient and train them to recognize new types of problems such as cancer or amyloid plaques. White blood cells called T cells and B cells are the major tool used by the body to defeat parasites, viruses, or toxins. They recognize these invaders because they have randomly created antibodies or receptor proteins on their surfaces that are able to dock onto foreign molecules; when this happens, they summon immune cells to break them down or destroy them. One therapeutic strategy that seems very promising is to remove T cells from a patient, grow them in the laboratory, and then equip them with receptors that can recognize unusual proteins that might be found on the surface of cancer cells, dangerous amyloid plaques, or other disease molecules. If this works, it might be possible to teach the body to confront cancer, degenerative diseases, and possibly even conditions such as aging in the same way that it fights infections.

Francis Collins (1950–   ), an American geneticist who heads the National Institutes of Health and led the Human Genome Project, recently discussed how he expects the relationship between genetics and medicine to evolve over the next decades. Some of his predictions, which appear on the MSNBC Web site (see the Further Resources section in the back matter), include the following:

- By 2010, tests will be developed to screen patients for genes linked to common diseases such as colon cancer, and over the next decade several types of gene therapy will be proven successful.

- By 2020, doctors will be using designer drugs to treat conditions such as diabetes and high blood pressure; therapies for cancer will have been developed based on molecules found in tumors, and it will be possible to diagnose and treat a number of genetically based mental illnesses.
- By 2030, researchers will have a list of genes involved in aging and will be carrying out clinical trials to extend people's lives; computer simulations will replace many types of laboratory experiments, and it will be common to sequence individual genomes.
- By 2040, medicine will have become individualized based on people's genetic profiles; in many cases molecular testing will warn doctors that there is a problem in advance of the appearance of disease symptoms, and gene therapy will be available for the treatment of most diseases.

Given the increasing pace of discoveries and technological developments, no one will be surprised if some of these milestones are reached earlier. On the other hand, it is entirely possible that scientists will discover new aspects of living systems that change the way they think about some neurodegenerative diseases, cancer, or other health problems. This may reveal that the diseases are much harder to treat. Yet it may also reveal ways of coping with them that are much simpler.

CONSCIOUSNESS AND THE BRAIN

"It's a scandal that science leaves out consciousness," said Christof Koch (1956–), a neuroscientist at the California Institute of Technology, in a 2006 interview with Caltech's Institute of International Studies. "Ten or twenty years ago, when we started, many scientists, probably the majority, said, 'Well, consciousness: we've got to leave that to the religious people, we've got to leave that to the philosophers, we've got to leave that to the New Age cult. That's not something scientists can study.' But that's silly. We are conscious and I believe it's the most essential aspect of my life, it's the fact that I'm a conscious being, and if I leave that out, then I will forever deprive science of one of the key aspects of the natural world."

Today's biology is a materialist science; it aims to explain things in physical terms. As Koch points out, for centuries philosophers, religious thinkers, and scientists have considered consciousness inexplicable, even off limits to materialist investigations. But no science of the human brain can be complete or satisfying unless it can explain what people usually regard as its most interesting feature, and Koch believes that neurobiology is ready to take on the theme. Over the course of 20 years, he pursued the topic with Francis Crick, codiscoverer of the structure of DNA and a founding father of molecular biology. Long before his death in 2004, Crick called consciousness the "major unsolved problem in biology," and began a quest to discover how the biology of the brain could produce this unique phenomenon. He found an excellent sounding board, critic, and partner in Koch, who continues to pursue the question in his laboratory at Caltech. The aim, he says, is ultimately to understand how a physical system like the brain can feel things—pain or pleasure, the sense of being angry, and self-awareness.

The researchers settled on a unique approach to the problem. Koch points out that while consciousness is an integral part of some human activities, others are done without conscious control. People digest food, ride bikes, and even have conversations without having to plan every sentence deliberately. Suddenly a person finds himself sitting in the car in the driveway, with no memory of a drive home from work, because his mind has been on something else. Koch calls such automatic activity zombie agents and says that life would not be possible without them. Some animals might have a sort of consciousness, he says—especially complex ones—but in other species, zombie agents might be able to manage all the activities they need to survive and reproduce. Realizing that zombie agents exist in humans permits scientists to look at how the brain manages them, and then the goal is to discover what makes them different from consciousness.

An important concept in Koch's work is the idea of a "neuronal correlate of consciousness" (NCC). He defines it as the "minimal set of mechanisms in your head that you need in order to be conscious." It is likely to be much smaller than the entire brain, because consciousness can operate when entire parts of the system are inactive. "I can close my eyes and I can visualize, so I don't need my eyes. Do I need my cerebellum? Probably not for visual consciousness or . . . any consciousness. So you can ask the question, what are the minimal set of mechanisms in your head that you need in order to be conscious? Is there a specific neurosignature, are there specific types of neurons, is there a particular type of neural activity, are there particular types of molecules, particular types of synapses, do they sit in a particular part of the brain? . . . Can you track them, can you catch a brief picture of them using some fancy imaging technique? Can you influence them?"

Most of the work of Koch's lab has focused on visual consciousness because it is easy to set up experiments to test what people see and what they do not. Magic tricks are often successful because the performer successfully directs people's attention away from what he does not want them to see—even when an object or a movement is plainly visible. The same effect can be achieved with a test subject looking at a video screen. By monitoring brain activity with an imaging technique such as MRI, the researchers look for areas of the brain that light up when a person becomes conscious of something. "It's not going to be one area, we know that," he says, "it's going to be a series of areas, probably distributed, that have different properties."

In this view, consciousness cannot be pinned down to a specific region of the brain or set of neurons. Koch says it is more like a "coalition of neurons, a little bit like in a democracy where you have coalitions that form, and that assemble and then disassemble . . . For a hundredth of a millisecond you may have this coalition of 5 million neurons. . . . They may give rise to a feeling of 'darn it, I'm late today,' and then this is suppressed because then there's this other 5 million neurons, or 10 million

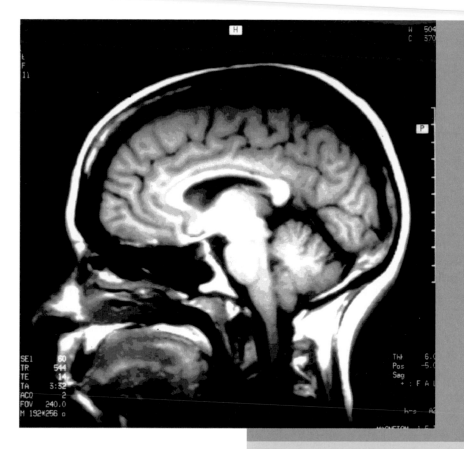

MRIs reveal the flow of blood through the brain as it performs various functions. This technique has been important in allowing researchers to determine which regions of the brain are crucial to various types of activity, as well to assess damage that has occurred through injuries, strokes, and other diseases. *(Dan S. Heffez)*

neurons, who now give rise to the 'oh, I see my daughter over there.'"

The coalitions compete for the attention of the host. Finding the NCC will involve looking at a brain pattern, tracking the formation of coalitions, and learning to recognize which ones lead to consciousness. If Koch and his colleagues can get a grip on visual processing, they hope that the same principles will apply to the way the brain manages awareness of other types of sensory information. Then it should be possible to look for telltale signs of consciousness in a fetus, a person in a coma, a dog, or a fly.

The scientists have already established that the frontal lobe—a higher part of the brain that arose in mammals, relatively recent in evolution—seems to be actively involved in directing a person's attention to specific parts of a stimulus. One amusing example that Koch has frequently used in lectures challenges the audience to watch a film of a group of six people, moving around and tossing two basketballs to each other. The task is to count the number of times the basketballs are tossed. Doing so requires such concentration that very few people in the audience notice that an actor in a gorilla suit walks leisurely by, stops and beats his chest, then moves on. He is perfectly obvious to anyone who is not busy counting. To Koch and his colleagues, this shows that the brain does not simply passively review visual information coming in through the eyes and assemble it into a story—instead, it helps direct attention and awareness through feedback loops that tell the eyes what to look for.

While Koch obviously hopes that the search for the signature of consciousness will be successful, he admits that there is no guarantee. Just as there is no chance that an ant can understand the theory of relativity, people might not have the mental capacity to understand consciousness. On the other hand, he feels that the present state of technology makes it worthwhile to pose the questions. His laboratory is now investigating whether there are specific types of neurons devoted to consciousness, to determine whether specific parts of the forebrain are required for consciousness, and to map the routes of feedback mechanisms between the frontal lobe and other parts of the brain.

Some answers to these questions may come from studies of the brains of people as they enter different conscious states—deep, unconscious sleep and dreaming; more may come from studies of people whose mental life has been disturbed because of an injury or a disease. One fan of the work of Koch and Crick is Oliver Sacks, a neurologist at the Albert Einstein College of Medicine in New York. Sacks is a brilliant observer who has written a number of insightful books on the way people adapt when things begin to go wrong in their brains. He deeply empathizes with these people and sees their conditions as gateways to understanding questions about the mind and consciousness.

In 2005, following a talk by Koch at the New York Academy of Sciences, Sacks commented: "As a clinician, I see patients with problems. They can be thought of as experiments of nature . . . Pathologies and illusions are a wonderful subject for examining the connections between the mind and the nervous system."

THE QUEST FOR ETERNAL YOUTH AND IMMORTALITY

In 1993, researchers in the laboratory of the biochemist Cynthia Kenyon at the University of California, San Francisco, discovered that a change in a single gene doubled the life span of a laboratory organism, a small worm called *Caenorhabditis elegans.* It was a startling finding. "These mutant worms still looked and acted young when they should be old," Kenyon wrote on her laboratory Web site (see the Further Resources section in the back matter). "Seeing them was like talking to someone that looks 40 and learning that they were really 80. This was a stunning finding because no one thought it was possible." Further work on the worm, mice, and human cells suggests that similar mechanisms may control the life span and the process of aging throughout the animal kingdom. The discovery has triggered a new way of thinking about aging, and serious efforts across the world to find ways to cure it.

Kenyon had long wondered whether genes influenced the process of aging—an unconventional question. Most of her colleagues considered the deterioration of the body as simply a natural process: an accumulation of errors in cells, leading to damage in DNA that might cause cancer, defective proteins, and accumulations of junk such as amyloid plaques that eventually disrupted the functions of organs. Many thought that aging was like taking a photo of a painting and then photographing the photo, over and over, losing quality each time, until the image became unrecognizable. The immune system had not evolved mechanisms to prevent aging because natural selection has only a very weak effect on organisms once they pass the age of reproduction. Still, Kenyon thought that there had to be genetic controls on the mechanisms that controlled aging and an organism's life span.

Some hints that she might be right had appeared a few decades earlier. Until 1962, most researchers had believed that human cells grown in laboratory cultures could keep reproducing forever. In that year, Leonard Hayflick, a professor of research medicine at the Wistar Institute in Philadelphia, showed that they had a limited life span. Depending on the conditions in which they were grown, they reproduced themselves from 50 to 70 times and then the whole population died. Part of the reason became clear in the 1970s with the discovery of *telomeres,* unusual regions of DNA at the end of chromosomes.

The postdoctoral fellow Elizabeth Blackburn, working at Yale University in the laboratory of the cell biologist Joseph Gall, was investigating a curious phenomenon related to cell division. DNA is copied by molecular machines, but they have a limitation: They cannot copy all the way to the ends of chromosomes. This means that each time a cell divides, they lose a bit of information at the tips. If there were genes in these regions, this process would take larger and larger bites out of them, quickly destroying key parts of the molecules. Blackburn and her colleagues found that evolution had provided a solution: long DNA sequences that did not contain genes had evolved at the tips of chromosomes. Sequences are still lost with every cell division, but a lot of junk has to be carried away before any genes are affected. This acts like a timer that allows the cell to use all of its genes until the telomere is gone.

Blackburn and Gall also discovered that the timer was sometimes turned back a bit to extend the cell's life span. Some types of cells made proteins called *telomerases* that added new DNA to the telomeres. In a speech given as Blackburn accepted the 1998 Australian Prime Minister's Science Prize, she compared the process to shoelaces: "If you don't have those little tips on both ends of your shoelace, the shoelace frays," she said. "Even worse, without telomeres, broken chromosome ends combine with any other end they find and that is not good for the health of the organism. It's as though someone ties your shoe laces together and makes you fall over." Telomerases allow stem cells, embryonic cells, and a few other types to divide more times because they produce telomerases. But that production stops in most adult cells, and the timer begins its countdown toward aging and eventual death. One way to extend life might be by fooling more types of cells into making telomerases and to have them keep doing so for a long time. On the other hand, the effects of the molecules are not always positive. Cancer cells sometimes use the same trick to overcome the cell division timer. So telomeres and telomerases are of interest to cancer researchers as well as those working on aging.

Such findings supported Cynthia Kenyon's feeling that genetic mechanisms might have something to say about life spans and aging. "After all, rats live three years and squirrels can live for twenty-five, and these animals are different because of their genes. Also, most biological processes are subject to tight control by the genes. If so, then by finding genes that control aging, and then changing the activities of the proteins they encode, one day we might be able to stay young much longer than we do now."

In the early 1990s, the worm *C. elegans* was becoming a favorite of scientists. Sydney Brenner (1927–), was using it in his laboratory at the Medical Research Council in Cambridge, Britain. Mutations in genes had immediate, obvious effects on the worm's body plan, and a series of discoveries about animal development by Brenner's lab earned him a 2002 Nobel Prize in physiology or medicine. Kenyon had worked with him as a postdoctoral fellow. When she got an independent position at UCSF in the mid-1980s, she brought the organism along, intending to use it to study aging. In 1993, Ramon Tabtiang, a student in her lab, discovered mutations that doubled the life span of *C. elegans.* Normally, the worms live about 21 days, but mutations in a gene called daf-2 produced animals that lived about 45 days—and they remained active and healthy to the end.

Interestingly, the gene has an important role in the biology of the worm: When faced with overcrowding or starvation, its larvae go into a sort of holding pattern called the Dauer state. They stop developing and aging and survive four to eight times longer than their counterparts. If food becomes available again, the worms complete their development into adults. Daf-2 plays an important role in switching this pause condition on and off. The discovery that the same gene extended the worms' life span—without triggering the Dauer state—convinced Kenyon that it was part of a more general life extension mechanism.

The group identified a second gene, called daf-16, which also contributed to keeping the worms young. Since these discoveries, Kenyon's lab and others have unraveled some of the reasons why. "We now know that these genes, daf-2 and daf-16, allow the tissues to respond to hormones that affect life span. We showed that daf-2 and daf-16 ultimately affect life span by influencing the activities of a wide variety of subordinate genes that influence the level of the body's antioxidants, the power of its immune system, its ability to repair its proteins, and many other beneficial processes. . . . This knowledge has now allowed us to extend the life span of active, youthful worms by sixfold."

One conclusion from the work has been to demonstrate that life span and aging are not necessarily tightly bound to each other. Few people would choose to double their life span if it meant living for another century with Alzheimer's disease or if the body underwent more and more severe deterioration. But as Kenyon writes, "Especially wonderful is the fact that these long-lived animals are resistant to a variety of age related diseases, including (in various animals) cancer, heart failure, and protein-aggregation disease. Thus these mutants not only look young, they are young, in the sense that they are not susceptible to age-related disease until later. . . . This link between aging and age-related disease suggests an entirely new way to combat many diseases all at once; namely, by going after their greatest risk factor: aging itself. This is an extremely exciting and important concept that could revolutionize medicine, human health and longevity, and it has just now begun to be studied in earnest, still in only a handful of labs."

Can Kenyon's results be extended to humans and possibly turned into a method of extending the length and quality of life? Many researchers are convinced that they can be, at least to some extent, especially since the discovery that the genomes of animals ranging from flies to mice to humans contain molecules related to daf-2—and they have similar functions. In the worm, daf-2 sits on the surfaces of cells, where it senses hormones. These are small molecules that carry signals through the body, helping it adapt to changes—sometimes sudden ones. In humans, the hormone insulin, for example, acts as a monitoring device that helps the body adjust to the presence or absence of food. It travels through the bloodstream, docks onto receptor proteins on cells, and triggers the activation of genes. *C. elegans* also produces an insulin-like molecule, and the molecule it docks onto is daf-2.

Where the worm has one molecule, evolution has given human beings two. The closest relatives of daf-2 in people are insulin and another hormone called the insulin-like growth factor 1 (IGF-1). To Kenyon and a number of other researchers, this suggests that the human receptors might also have played a role in aging. Interestingly, it would also provide a connection between that process and a person's diet.

Researchers across the world have been investigating this question in mice and other laboratory animals. It is one theme being pursued at the Italian station of the European Molecular Biology Laboratory near Rome, in the group of the developmental biologist Nadia Rosenthal (who wrote the Foreword to this book). A main focus of Rosenthal's work is muscle, particularly the heart, and diseases related to muscle development and deterioration over the course of a lifetime. IGF-1 drew her attention in the 1990s, while

Nadia Rosenthal, an American geneticist with colleagues in Italy and Australia, is author of each foreword in this multivolume set. Her research is providing insights into the molecular signals that prompt the regeneration of tissues by stem cells and genes involved in the process of aging. (*Nadia Rosenthal*)

she was working at Harvard University, because the hormone was thought to play an important role in triggering the formation of muscle cells. Normally, the signal is active in embryos and cases where damaged muscle needs to be repaired.

Rosenthal wondered what would happen if adult muscles could produce the growth factor themselves. In collaboration with H. Lee Sweeney's lab at the University of Pennsylvania, her group developed a strain of mouse in which particular muscle cells produced IGF-1 locally throughout their entire life spans. The researchers discovered that the factor seemed to be activating a "regenerative program" that could recruit stem cells to form new muscle very efficiently. The mice were healthy and so muscular that lab technicians gave them the nickname "Schwarzenegger mouse," and they remained amazingly fit even at the "advanced age" of 20 months—the mouse equivalent of retirement age. IGF-1 was holding off the normal deterioration of muscle and helping to rebuild it in mice that had already lost muscle mass. And it was significantly increasing the animals' health span. In the case of the Schwarzenegger mouse, the local production of IGF-1 was acting to protect the tissue environment from the deterioration of age by inducing new signals in the muscle.

This is only one example of an enormous amount of ongoing work that has established a connection between the body's hormone systems—which are closely linked to diet—and aging. Insulin and IGF-1 provoke different responses in different tissues. While both molecules are vital to growth and the way the body processes food, Kenyon and many others believe that keeping insulin levels low is generally good for animals. "What's really interesting is that you can get the life span benefits by taking away the insulin receptor in individual tissues," she says. "So it might not be overall percentage of insulin function we need to concentrate on, but a selective percentage in different tissues—like fat cells."

Kenyon has formed a company called Elixir Pharmaceuticals that aims to develop therapies based on manipulating the body's response to insulin. A cure for aging is not anywhere near on the horizon. But many scientists now believe that genes may be a key to lengthening life and improving its qual-

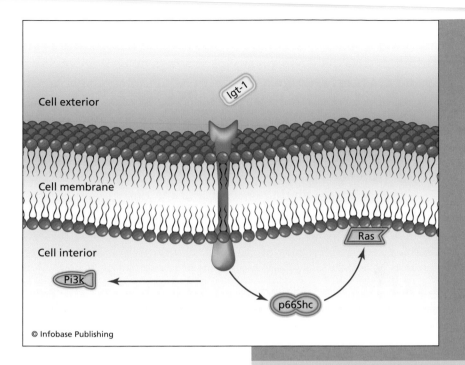

Cell exterior

Igt-1

Cell membrane

Cell interior

Pi3k

Ras

p66Shc

© Infobase Publishing

The IGF-1 receptor is embedded in the surface of cells and responds to an insulin-like hormone. Researchers have shown that its activity is related to aging.

ity in old age. In an interview for the journal *PLoS Biology*, Kenyon put the issue into an evolutionary perspective. Hundreds of millions of years ago, humans' ancestors were also small worms. "If you'd asked me . . . 'Cynthia, you have a two-week lifespan, do you think that you could [live longer]?' And if I'd told you, 'Well, I think our descendants will live 1,000 times longer,' you'd have said, 'Oh, come on!' But we do. It happened."

SAVING THE WORLD: ANTS, HUMANS, AND THE FATE OF THE EARTH

It might seem strange that a scientist whose work has mainly focused on a tiny insect spends a great deal of his time considering the fate of the entire planet. But Edward O. Wilson (1929–   ), a professor at Harvard and the world's foremost expert on ants, has always had a much larger perspective on life. He has been awarded two Pulitzer Prizes for his writings on ecology and the natural sciences. Part of his broader view of the world has come from witnessing the increasingly rapid destruction of habitats occupied by ants and many other species—mostly as a result of human activity. Wilson believes that the situation is dire, and in 2002 he collected his thoughts in a book called *The Future of Life*—a vision of what the world will be like if present trends such as human population growth and the destruction of ecological systems continue.

Thomas Malthus (1766–1834), a British scholar and pastor, seems to have been one of the first people to understand that overpopulation could lead to disaster. He was the first to describe a connection between the growth of populations—which puts stress on the food supply and the environment—and poverty. This relationship was so misunderstood in the late 18th century that the British government was providing subsidies that encouraged the poor to have more children. As a result, the country's population was growing at an alarming rate.

"The power of population is indefinitely greater than the power in the Earth to produce subsistence for man," Malthus wrote in the first edition of *An Essay on the Principle of Population,* published in 1798. The message was that every new human child has the potential to go on to create many new mouths to feed, whereas the surface of the Earth does not inflate as human populations expand. It has finite resources, and the amount that it can produce grows much more slowly. Malthus also saw the connection between overpopulation, wars, and epidemics. When Charles Darwin and Alfred Russel Wallace read Malthus's essay, they began thinking about how nature kept species in check—for example, why the surface of the world was not a towering pile of ants that rose miles into the air—and the result was the theory of evolution.

Evolution and ecological thinking arose hand in hand, because the theory showed how dependent species are upon one another. In some cases, these interdependencies are obvious—if one species is wiped out by a disease, those that feed on it will also suffer. But often the relationships are more subtle, based on networks of interactions between many species that are difficult to uncover. Those must be understood, Wilson believes, because only then will people be able to perceive the dangerous effects that their own behavior and lifestyle are having on the planet as a whole.

The Future of Life opens with some startling facts. For example, Wilson describes a person's "ecological footprint—the average amount of productive land and shallow sea appropriated by each person in bits and pieces from around the world for food, water, housing, energy, transportation, commerce, and waste absorption." In developing nations, an individual's footprint is currently about 2.5 acres, whereas in the United States it is nearly 10 times as large (24 acres). Bringing everyone on Earth to this level, he writes, "would require four more planet Earths. The 5 billion people of the developing countries may never wish to attain this level of profligacy. But in trying to achieve at least a decent standard of living, they have joined the industrial world in erasing the last of the natural environments."

This erasure involves the clearing of land to build new homes and cities, of course, but there are many other factors. Conservationists summarize the reasons for the decline of species with the acronym HIPPO, which stands for habitat destruction, invasive species, pollution, population, and overharvesting. For the first time, metagenomics methods (described in the previous chapter) are allowing scientists to determine just how bad the damage really is. Even if the situation proves not as serious on the microbial scale as it is for larger species, whose numbers are easier to measure, Wilson says that the worldwide situation is already extremely serious, as can be seen through the example of Hawaii, which he calls a laboratory in which it is possible to understand what is happening throughout the rest of the world.

Hawaii is so distant from the nearest major landmass that it took a long time for other species to settle there. Wilson and his colleagues estimate that on average, one new species may have arrived every 1,000 years, carried by winds or floating on bits of wood. "Extremely few made a successful landfall. Even then the pioneers faced formidable obstacles. There had to be a niche to fill immediately upon arrival—the right place to live, the right food to eat, potential mates immigrating with them, and few or no predators waiting to gobble them up." Once a species had settled in, it began to adapt into forms unique to the islands.

Settling Hawaii took a long time; killing off its species is going much more quickly. Originally, Hawaii was home to more than 125 species of birds that existed nowhere else; only 35 remain, 24 of which are considered endangered. The island's unique plants and insects face a similar situation. Humans have imported a wide range of other species—either deliberately for food or other uses or by accident—that have wiped out the native species. This sort of replacement happens naturally, of course, but with the help of humans, what used to take millions of years now happens in decades. Three-quarters of Hawaii's land has been converted into living space or fields for crops. Humans brought along pigs, some of which escaped and became wild predators. But perhaps the greatest blow to the environment, Wilson says, is one of the tiniest threats—ants. Hawaii never had them, so its species never had to adapt to them. Their arrival with humans caused a huge shock to Hawaii's system.

Once a unique species is lost, it is lost forever, barring a rapid jump forward in cloning technology and a massive effort to revive a species. There are practical reasons to be concerned about these losses: Many of today's most potent drugs to fight cancer and other diseases have come from exotic sources such as peculiar fungi found on the bark of trees. But the real issue is a much deeper one. "Earth, unlike other solar planets, is not in physical equilibrium," Wilson writes. "It depends on its living shell to create the special conditions on which life is sustainable. The soil, water, and atmosphere of its surface have

(opposite page) The Hawaiian Islands are so distant from the continental mainland that they were settled very slowly by foreign species, which have evolved in unique ways over millions of years. When humans arrived they brought along species that have been wiping out many of these unique plants and animals at a rapid pace.

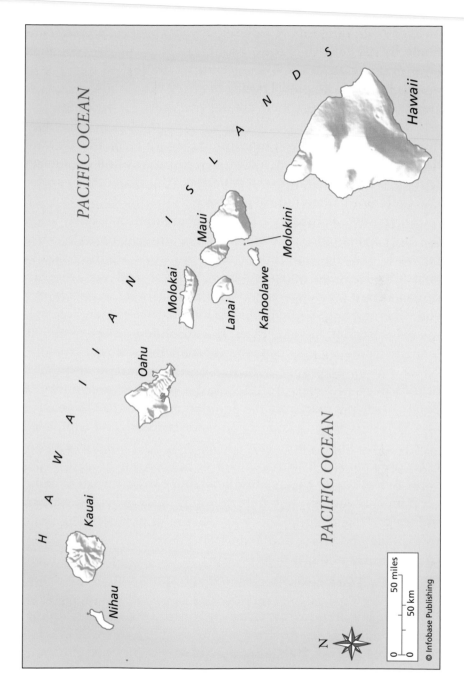

PACIFIC OCEAN

PACIFIC OCEAN

H A W A I I A N I S L A N D S

Kauai

Nihau

Oahu

Molokai

Lanai

Maui

Kahoolawe

Molokini

Hawaii

N

0   50 miles

0   50 km

evolved over hundreds of millions of years to their present condition by the activity of the biosphere, a stupendously complex layer of living creatures whose activities are locked together in precise but tenuous global cycles of energy and transformed organic matter. The biosphere creates our special world anew every day, every minute. . . . When we alter the biosphere in any direction, we move the environment away from the delicate dance of biology. When we destroy ecosystems and extinguish species, we degrade the greatest heritage this planet has to offer and thereby threaten our own existence."

Of the five elements of HIPPO, population has likely been the most serious and will continue to be, affecting both the natural world and the evolution of society. Just as Malthus discovered links between overpopulation, disease, and war, modern ecologists see connections between the birthrate in developing countries and huge social problems such as government instability and terrorism. In some cases, the absolute number of people living in a country may be less important than its age structure, which has an enormous influence on the economy and many other aspects of society. In 1999, in the West African country of Benin, nearly 50 percent of the population was under 15 years of age. Wilson writes, "A country poor to start with and composed largely of young children and adolescents is strained to provide even minimal health services and education for its people. Its superabundance of cheap, unskilled labor can be turned to some economic advantage but unfortunately also provides cannon fodder for ethnic strife and war. As the populations continue to explode and water and arable land grow scarcer, the industrial countries will feel their pressure in the form of many more desperate immigrants and the risk of spreading international terrorism."

There is a limit to how many people the planet can support. If everyone ate grain, the Earth might be able to provide enough for a population of 10 billion; if instead, people used the grain to feed animals to provide meat, Wilson estimates the number might only be 2.5 billion. But food production is coupled to many other things, particularly the need for water. Wilson explains: "A thousand tons of freshwater yields a ton of wheat, worth $200, but the same amount of water in industry yields $14,000." If the main interest is profit, this means that water will inevitably be diverted from food production to industry. This leads to pollution, which makes prices rise higher because the water will have to be treated and purified before it can be returned to agriculture. Industrial countries may be able to afford the exorbitant prices, but others will not.

Wilson is not a doomsayer who stops with a critique of the current situation; he proposes a number of concrete measures that could have a significant impact over the long term, including:

- Set aside and stringently protect the natural hot spots with the greatest biodiversity that are currently most at risk. Wilson estimates that 25 percent of these ecosystems take up only 1.4 percent of the Earth's land, but are home to 43.8 percent of the vascular plants and 35.6 percent of known animal species.
- Keep the rain forests of the Amazon and four other remaining frontier forests intact
- Cease all logging of old-growth forests everywhere
- Protect river and lake systems everywhere
- Identify the most important marine hot spots, such as coral reefs, and protect them
- Obtain as complete a map as possible of the Earth's biodiversity
- Make conservation profitable by helping people who live near preserves, supporting them financially and involving them professionally in conservation efforts
- Investigate biodiversity more thoroughly in order to use more species as sources of food and pharmaceuticals and to take advantage of species' ability to restore the environment after it has been damaged
- Develop genetically modified crops that have been thoroughly tested for safety and use them in a regulated way
- Increase the capacity of zoos and botanical gardens to breed endangered species
- Support population planning throughout the world

If these measures can be taken quickly, as a collaboration between governments, scientists, and industries, Wilson believes
that it is not too late to reverse the major damage that has been
done. It should be possible to feed humanity, if populations
can be stabilized by the middle of the 21st century, and if the
world's ecospheres are protected and restored.

Modern conservation provides a sobering perspective to
those who have believed that science will progress fast enough
to solve all the world's problems and ensure a healthy future for
humanity. Evolution, genetics, and the other parts of modern
biology have demonstrated that human nature is dependent on
both genes and the environment. Human lifestyles have already
had a deep impact on ecospheres; increasingly and inevitably,
these changes will mold the future of the species. One of the
most important lessons of modern biology is that the only way
to provide a healthy future for humanity is to improve the quality of life for people around the globe—using the tools of genetics and other means—while ensuring that the Earth is a healthy
place to live.

Chronology

1651   William Harvey claims that all animals arise from eggs.

1677   Antoni van Leeuwenhoek discovers sperm.

1751   Pierre-Louis Maupertuis studies polydactyly, the inheritance of extra fingers in humans.

              Joseph Adams recognizes the negative hereditary effects of inbreeding.

1802   Jean-Baptiste Lamarck publishes *Research on the Organization of Living Bodies,* in which he claims that species become more perfect and pass on acquired characteristics to their offspring. The hypothesis is overturned by evolution.

1824   Joseph Lister builds a new type of microscope that removes distortion and greatly increases resolution.

1838   Matthias Schleiden discovers that plants are made of cells.

1840   Theodor Schwann discovers that all animal tissues are made of cells.

1856   Gregor Mendel begins experiments on heredity in pea plants.

1857	Joseph von Gerlach discovers a new way of staining cells that reveals their internal structures.
1858	The theory of evolution is made public at a meeting of the Linnean Society in London with the reading of papers by Charles Darwin and Alfred Russel Wallace.
	Rudolf Virchow states the principle of *Omnis cellula e cellula*: every cell derives from another cell—including cancer cells.
1859	Charles Darwin publishes *On the Origin of Species*. The complete first print sells out on the first day.
1865	Gregor Mendel presents his paper "Experiments in Plant Hybridization" in meetings of the Society for the Study of Natural Sciences in Brnø, Moravia. The paper outlines the basic principles of the modern science of genetics. It is published the next year but receives little attention.
1868	Fredrich Miescher isolates DNA from the nuclei of cells; he calls it nuclein.
1871	Francis Galton carries out experiments in rabbits that disprove Darwin's hypothesis of how heredity functions.
1876	Oscar Hertwig observes the fusion of sperm and egg nuclei during fertilization.

1879	Walther Flemming observes the behavior of chromosomes during cell division.
1885	August Weismann states that organisms separate reproductive cells from the rest of their bodies, which helps explain why Lamarck's concept of evolution and inheritance is wrong. He tries and fails to observe Lamarckian inheritance in the laboratory by cutting off the tails of mice for many generations.
1900	Hugo de Vries, Carl Correns, and Erich von Tschermak-Seysenegg independently publish papers that confirm Mendel's principles of heredity in a wide range of plants.
	Archibald Garrod identifies the first disease that is inherited according to Mendelian laws, which means that it is caused by a defective gene.
	Theodor Boveri demonstrates that different chromosomes are responsible for different hereditary characteristics.
1901	Karl Landsteiner identifies the ABO blood groups.
1902	William Bateson popularizes Mendel's work in a book called *Mendel's Principles of Heredity: A Defense*.
1903	Walter Sutton connects chromosome pairs to hereditary behavior, demonstrating that genes are located on chromosomes.

1905	Nettie Stevens and Edmund Wilson independently discover the role of the X and Y chromosomes in determining the sex of animal species.
1906	William Bateson discovers that some characteristics of plants depend on the activity of two genes.
1908	Archibald Garrod shows that humans with an inherited disease are lacking an enzyme (a protein), demonstrating that there is a connection between genes and proteins.
1909	William Bateson coins the term *genetics.*
1910	The Eugenics Record Office is opened at Cold Spring Harbor, New York.
	Thomas Hunt Morgan discovers the first mutations in fruit flies, *Drosophila melanogaster,* bred in the laboratory. This leads to the discovery of hundreds of new genes over the next decades.
1911	Morgan discovers some traits that are passed along in a sex-dependent manner and proposes that this happens because the genes are located on sex chromosomes. He proposes the general hypothesis that traits that are likely to be inherited together are located on the same chromosome.
1913	Alfred Sturtevant constructs the first genetic linkage map, allowing researchers to

pinpoint the physical locations of genes on chromosomes.

1920   Hans Spemann and Hilde Proescholdt Mangold begin a series of experiments in which they transplant embryonic tissue from one species to another. The scientists show that particular groups of cells they called organizers send instructions to neighboring cells, changing their developmental fates.

1921   Erwin Baur, Eugen Fischer, and Fritz Lenz publish a book called *Menschliche Erblichkeitslehre und Rassenhygiene,* which attempts to link genetics to race and is used by eugenicists in the United States and Germany as a justification for declaring that there are inferior races and a motivation for sterilizing and killing social undesirables.

1922   Ronald A. Fisher uses mathematics to show that Mendelian inheritance and evolution are compatible.

1927   Hermann Muller shows that radiation causes mutations in genes that can be passed down through heredity.

1928   Fredrick Griffith discovers that genetic information can be transferred from one bacterium to another, hinting that hereditary information is contained in DNA.

1931   Barbara McClintock shows that as chromosome pairs line up beside each other during the copying of DNA, fragments can break off one chromosome and be inserted into the other in a process called recombination.

Archibald Garrold proposes that diseases can be caused by a person's unique chemistry—in other words, genetic diseases may be linked to defects in enzymes.

1933   Theophilus Painter discovers that staining giant salivary chromosomes in fruit flies reveal regular striped bands.

1934   Calvin Bridges shows that chromosome bands can be used to pinpoint the exact locations of genes.

1935   Nikolai Timofeeff-Ressovsky, K. Zimmer, and Max Delbrück publish a groundbreaking work on the structure of genes that proposes that mutations alter the chemistry and structure of molecules.

1937   George Beadle and Boris Ephrussi show that genes work together in a specific order to produce some features of fruit flies.

1940   George Beadle and Edward Tatum prove that a mutation in a mold destroys an enzyme and that this characteristic is inherited in a Mendelian way, leading to their hypothesis that one gene is related to one enzyme (protein), formally proposed in 1946.

1943   Max Delbruck and Salvador Luria demonstrate evolution in the laboratory by showing that bacteria evolve defenses to viruses through mutations that are acted on by natural selection.

1944   Oswald Avery, Colin MacLeod, and Maclyn McCarty show that genes are made of DNA.

Erwin Schrödinger publishes *What Is Life?*

1948   The American Society for Human Genetics is founded.

1950   Barbara McClintock publishes evidence that genes can move to different positions as chromosomes are copied.

Erwin Chargaff discovers that the proportions of A and T bases in an organism's DNA are identical, as are the proportion of Gs to Cs.

1951   Rosalind Franklin uses X-ray diffraction to obtain images of DNA; the patterns reveal important clues to the building plan of the molecule.

1953   James Watson and Francis Crick publish the double helix model of DNA, which explains both how the molecule can be copied and how mutations might arise.

In the same issue of the journal *Nature,* Rosalind Franklin and Maurice Wilkins publish X-ray studies that support the Watson-Crick model. This launches the field of molecular biology that shows, over the next 20 years, how the information in genes is used to build organisms.

1958   Francis Crick describes the central dogma of molecular biology: DNA creates RNA creates proteins. He challenges the scientific community to figure out the molecules and mechanisms by which this happens.

1959   Jerome Lejeune discovers the first disease due to defects in chromosomes: Down syndrome is caused by the inheritance of an extra chromosome.

Marshall Nirenberg, Marianne Grunberg-Manago, and Severo Ochoa show that the cell reads DNA in three-letter words to translate the alphabet of DNA into the 20-letter alphabet of proteins.

1961   Sidney Brenner, François Jacob, and Matthew Meselson discover that messenger RNA is the template molecule that carries information from genes into protein form. Crick and Brenner suggest that proteins are made by reading three-letter codons in RNA sequences, which represent three-letter codes in DNA. M. W. Nirenberg and J. H. Matthaei use artificial RNAs to create proteins with specific spellings, helping them learn the complete codon spellings of amino acids.

1965   Leonard Hayflick discovers that human cells raised in laboratory cultures have a limited life span, prompting a search for molecular mechanisms of aging.

1966	Marshall Nirenberg and H. Gobind Khorana work out the complete genetic code—the DNA recipe for every amino acid.
1970	Hamilton Smith and Kent Wilcox isolate the first restriction enzyme, a molecule that cuts DNA at a specific sequence—which will become an essential tool in genetic engineering.
1972	Janet Mertz and Ron Davis use restriction enzymes and DNA-mending molecules called ligases to carry out the first recombination: the creation of an artificial DNA molecule. Paul Berg creates a new gene in bacteria using genetic engineering.
1973	Stanley Cohen, Annie Chang, Robert Helling, and Herbert Boyer create the first transgenic organism by putting an artificial chromosome into bacteria.
1975	Edward Southern creates Southern blotting, a method to detect a specific DNA sequence in a person's DNA; the method will become crucial to genetic testing and biology in general. Cesar Milsein, Georges Kohler, and Niels Kai Jerne develop a method to make monoclonal antibodies.
1977	Walter Gilbert and Allan Maxam develop a method to determine the sequence of a DNA molecule; Fredrick Sanger and colleagues

independently develop another very rapid method for doing so, launching the age of high-throughput DNA sequencing.

Frederick Sanger finishes the first genome, the complete nucleotide sequence of a bacteriophage.

Phillip Sharp and colleagues discover introns, information in the middle of genes which do not contain codes for proteins and must be removed before an RNA can be used to create a protein.

1977	Genentech, the first biotech firm, is founded based on plans to use genetic engineering to make drugs.
1978	Recombinant DNA technology is used to create the first human hormone.
1980	Christiane Nüsslein-Volhard and Eric Wieschaus discover the first patterning genes that influence the development of the fruit fly embryo, bringing together the fields of developmental biology and genetics.
1981	Three laboratories independently discover oncogenes: proteins that lead to cancer if they undergo mutations.
1982	Insulin becomes the first genetically engineered drug.
1983	Walter Gehring's laboratory in Basel and Matthew Scott and Amy Weiner, working

at the University of Indiana, independently discover HOX genes: master patterning molecules for the creation of the head-to-tail axis in animals as diverse as flies and humans.

1985   Kary B. Mullis publishes a paper describing the polymerase chain reaction, a method which rapidly and easily copies DNA molecules.

1986   First outbreak of BSE (mad cow disease) among cattle in the United Kingdom

1987   First human genetic map published

1988   The Human Genome Project is launched by the U.S. Department of Energy and the National Institutes of Health, with the aim of determining the complete sequence of human DNA.

1989   Alec Jeffreys discovers regions of DNA that undergo high numbers of mutations. He develops a method of DNA fingerprinting that can match DNA samples to the person they came from and can also be used in establishing paternity and other types of family relationships.

The Human Genome Organization (HUGO) is founded.

1990   W. French Anderson carries out the first human gene replacement therapy to treat an immune system disease in four-year-old Ashanti DeSilva.

1993   The company Monsanto develops and begins to market a genetically engineered strain of tomatoes called Flavr Savr.

            The Huntington disease gene is found.

            Cynthia Kenyon discovers mutations in *C. elegans* that double the worm's life span.

1994   Mary-Claire King discovers BRCA1, a gene that contributes to susceptibility to breast cancer.

1995   The first confirmed death from Creutzfeldt-Jakob disease, the human form of BSE, is reported in the United Kingdom.

1996   Researchers complete the first genome of a eucaryote, baker's yeast. The completion of the genome of *Methanococcus jannaschii,* an archaeal cell, confirms that archaea are a third branch of life, separate from bacteria and eucaryotes.

            Gene therapy trials to use the adenovirus as a vector for healthy genes are approved in the United States.

1997   Ian Wilmut's laboratory at the Roslin Institute produces Dolly the sheep, the first cloned mammal.

1998   Scientists obtain the first complete genome sequence of an animal, the worm *Caenorhabditis elegans.*

1999	Jesse Gelsinger dies in a gene therapy trial, bringing a temporary halt to all viral gene therapy trials in the United States.
2000	The genome of the fruit fly *Drosophila melanogaster* is completed.
	Scientists complete a working draft of the human genome. The complete genome is published in 2003.
2002	The mouse genome is completed.
2004	Scientists in Seoul announce the first successful cloning of a human being, a claim which is quickly proven to be false.
2008	Samuel Wood of the California company Stemagen successfully uses his own skin cells to produce clones, which survive five days.

Glossary

allele   one variant of a particular gene

amyloid plaque   a cluster of protein fragments that accumulates in tissues and does not dissolve, frequently found in association with Alzheimer's and other neurodegenerative diseases

apoptosis   a cellular self-destruct program triggered by genes, usually in response to external stimuli

biodiversity   coined from biological diversity, referring to the amount of life found in a particular environment

bioremediation   a process by which microorganisms restore features of the environment to their original state, for example when there has been contamination by pollutants or radiation

chromatin immunoprecipitation (ChIP)   procedure to determine whether a given protein binds to or is localized to a specific DNA sequence in vivo

clade   a branch of an evolutionary or family tree containing only organisms that have descended from a specific common ancestor

cloning   a method which makes an exact copy of a DNA sequence, a chromosome, or an entire genome

computerized tomography (CT)   a method which uses X-rays or another method to scan patient tissues, creating slicelike photographs that are assembled into three-dimensional images by computer

conditional mutagenesis   a type of genetic engineering that knocks out or knocks in a gene only in specific tissues, rather than in an entire organism, often in combination with molecules that allow researchers to decide when the alteration takes place

constitutional activation a form of a gene or molecule that is always active, usually because of mutations that remove its ability to be switched off

crystallography a method of turning proteins or other biological molecules into crystals, often the first step in determining a three-dimensional atomic structure of a molecule

DNA (deoxyribose nucleic acid) a molecule made of nucleic acids that forms a double helix in cells, holds a species' genetic information, and encodes RNAs and proteins

DNA microarray (DNA chip) a set of probes made of nucleic acids, usually mounted on a glass slide, used to compare the RNAs made by different types of cells

dominant an allele that determines the phenotype of an organism, even when that organism has a different allele as the second copy of the gene

eugenics strategies and actual programs to influence the gene pool and future evolution of humans by controlling their mating—in some cases, by sterilizing or killing people judged to be unfit

fermentation a process by which cells degrade substances to produce energy, in the absence of oxygen

fitness the degree to which an organism is adapted to its environment

forward genetics a method of discovering gene functions that starts with a phenotype and searches for the molecule that is responsible for it

gastrulation a process that takes place in the early development of the embryo, in which undifferentiated cells form three layers that go on to produce the body's major tissues and organs

gene a region of DNA that encodes a protein

genetically modified organism (GMO) an organism whose genes have been altered through an artificial process in the laboratory

genome the entire set of DNA in an organism or species, usually referring to the DNA in nucleus (of cells that have one)

genotype an organism's complete collection of genes, including both dominant and recessive alleles

germ cell a specialized cell capable of creating a new organism (a sperm or egg cell)

green fluorescent protein (GFP) a protein that releases green fluorescent light when exposed to energy of a particular frequency

heredity the means by which features of a parent organism are passed to its offspring

homologue a DNA sequence, tissue, organ, or other body structure that is the closest evolutionary relative of a similar structure in another organism

HOX gene a gene containing a structure called a homeobox, which usually controls important developmental processes in embryos

intron a sequence in an RNA or gene that does not encode a part of a protein

in vivo in the living body of an animal or plant

knock in a gene that has been artificially added to an organism

knock out a gene that has been artificially removed from an organism

ligase an enzyme that can join other molecules together; in genetic engineering, ligases are used to combine fragments of DNA to make new genes

lobotomy the surgical removal of part of the brain

magnetic resonance imaging (MRI)   a method that detects the presence and locations of particular atoms by exposing them to strong magnetic fields

mass spectrometry   a method of detecting the composition of substances, such as proteins or other chemical compounds

materialism   a philosophy that seeks physical and chemical explanations for phenomena, including mental behavior and states

messenger RNA   a molecule made of nucleic acids, based on the information in a gene, used as a template for the production of a protein

metagenomics   an approach to DNA sequencing that tries to capture the sequences of all DNA found in a particular environment rather than that of a particular organism or species

miasma theory of disease   an ancient idea that illness is caused by the inhalation of bad air

microRNA   a tiny RNA molecule produced by cells whose main function seems to be to bind to specific messenger RNAs and prevent their translation into proteins

microtubule   a cellular fiber made of protein subunits called tubulin that helps give the cell its shape and structure, participates in cell division, and serves as a highway along which other molecules are delivered

mitotic spindle   a structure built during cell division; it is made of microtubules, and its function is to separate chromosomes into two equal sets

monogenic trait   a feature of an organism that is determined by the presence of a particular allele of a single gene

mutagen   a substance that causes mutations in genes

mutation   a change in an organism's DNA sequence caused by a copying error, a mutagen, or some other form of damage

natural selection   the process by which the environment gives some members of a species an advantage at having more offspring due to their genetic makeup

nuclear magnetic resonance   a method that uses a very strong magnetic field to identify the atoms that make up molecules and plot their positions relative to each other, a common method of investigating the three-dimensional structures of proteins

nucleotide (base)   a subunit of DNA and RNA that consists of a base linked to a phosphate group and sugar

ontogeny   the stages of an individual organism's development, from fertilization to birth and adulthood

phenotype   the complete set of measurable physical and behavioral characteristics of an organism determined by its genes

phylogeny   the stages of a species's evolution, from the first cell to its current form

protein   a molecule made of amino acids, produced by a cell based on information in its genes

recapitulation   Ernst Haeckel's theory that the development of a single organism passes through phases that retrace its evolutionary history

receptor (protein)   a molecule in a cell or on its surface that binds to a specific partner molecule, usually leading to a change in its activity, and often ultimately resulting in a change in the set of genes active within a cell

recessive   an allele that must be present in two copies in an organism to fully determine its phenotype

restriction enzyme   a protein that cuts single-stranded or double-stranded DNA at specific sequences

reverse genetics   altering a gene to discover its functions in cells and observe its effects on an organism's phenotype

ribonucleic acid (RNA)   a molecule made of nucleotides that is produced by transcribing the information in a DNA sequence

sequence   a list of the subunits of a molecule such as DNA or proteins, in the order in which they are attached to each other

small interfering RNA (siRNA)   an artificial molecule made of RNA, used to prevent specific messenger RNAs from being translated into proteins

spontaneous generation   a disproven theory that held that complex living organisms such as maggots or flies commonly arose on their own without an egg

synchrotron   a circular instrument in which electrons or other subatomic particles are accelerated through the use of magnets, often used in biology as a source of high-energy X-rays to study the structures of molecules

systems biology   an interdisciplinary field that sees life as the result of complex networks of many interacting elements, usually attempting to study it through computer models

telomerase   an enzyme that adds small repeated DNA sequences to telomeres, helping protect chromosomes from being degraded

telomere   a region at the ends of chromosomes that does not contain genes but protects chromosomes from being degraded as DNA is copied

tiling array   a type of DNA microarray designed to investigate whether RNA molecules have been made from any of the sequences within a particular region of the genome, including segments that are not known to contain genes

tumor suppressor gene   a gene that leads to tumors when it becomes defective; the healthy form protects cells from becoming cancerous

variation   the diversity within a species, caused by the existence of different forms of genes, and the range of phenotypes that this diversity produces

viral therapy   methods that aim to use viruses to deliver healthy forms of genes or other genetic material to cells

vitalism   the hypothesis that a special form of energy (often thought to be spiritual or nonmaterial in nature) is necessary to produce life from nonliving substances and that life cannot be explained purely in terms of physical and chemical forces

Further Resources

Books and Articles

Allen, Myles. "A novel view of global warming." *Nature* 433 (January 2005). A climate researcher's critique of Michael Crichton's book *State of Fear,* which cites scientific publications to criticize the way the research community has dealt with the theme of global warming. Allen believes that Crichton distorted the evidence and is manipulating readers' opinions by injecting pseudo-science into a novel.

Asimov, Isaac. *The Beginning and the End.* New York: Pocket Books, 1983. Essays on genetics and other branches of science, with a particular focus on the impact they will have on humans and society in the future, from one of the world's foremost science fiction and popular science authors.

Bodman, Walter, and Robin McKie. *The Book of Man: The Quest to Discover Our Genetic Heritage.* London: Little, Brown and Company, 1994. A well-written account of the major people and themes of human genetics from the late 19th century to the beginning of the Human Genome Project.

Branden, Carl, and John Tooze. *Introduction to Protein Structure,* 2nd ed. New York: Garland Publishing, 1999. A detailed overview of the chemistry and physics of proteins, for university students with some background in both fields.

Brown, Andrew. *In the Beginning Was the Worm.* London: Pocket Books, 2004. The story of an unlikely model organism in biology: the worm *C. elegans,* and the scientists who have used it to understand some of the most fascinating issues in modern biology.

Browne, Janet. *Charles Darwin: The Power of Place.* New York: Knopf, 2002. The second volume of the definitive biography of Charles Darwin.

————. *Charles Darwin: Voyaging.* Princeton, N.J.: Princeton University Press, 1995. The first volume of the definitive biography of Charles Darwin.

Caporale, Lynn Helena. *Darwin in the Genome: Molecular Strategies in Biological Evolution.* New York: McGraw Hill, 2003. A new look at variation and natural selection based on discoveries from the genomes of humans and other species, written by a noted biochemist.

Carlson, Elof Axel. *Mendel's Legacy: The Origin of Classical Genetics.* Cold Spring Harbor, N.Y.: Cold Spring Harbor Laboratory Press, 2004. An excellent, easy-to-read history of genetics from Mendel's work to the 1950s. Carlson explains the relationship between cell biology and genetics especially well.

————. *The Unfit: A History of a Bad Idea.* Cold Spring Harbor, N.Y.: Cold Spring Harbor Laboratory Press, 2001. An in-depth account of eugenics movements across the world.

Carson, Rachel. *Silent Spring.* Cambridge: Riverside Press, 1962. Carson's book on the impact of DDT on birds and the rest of the ecosphere, proposing that chemical pollutants constitute a serious threat to the health of humans and other species. The book played an important role in the rise of environmental movements over the next decades.

Carter, Rita. *Mapping the Mind.* London: Phoenix, 2000. A beautifully written and illustrated book in which science writer Rita Carver explores the state of the art of research into higher levels of brain function. While the book does not deeply explore the genetics of the brain, it gives an excellent overview of what current methods have revealed about how brain tissues participate in perception, behavior, and other functions.

Cavalli-Sforza, L. Luca. "The Human Genome Diversity Project: Past, Present, and Future." *Nature Reviews Genetics* 6 (2005): 333. An overview of the progress and difficulties encountered by the HGDP from its conception to its development as a resource now being used by researchers all over the world.

Cavalli-Sforza, L. Luca, Paolo Menozzi, and Alberto Piazza. *The History and Geography of Human Genes.* Princeton, N.J.: Princeton University Press, 1994. For over three decades Cavalli-Sforza has been interested in using genes (as well as other fields such as linguistics) to study human diversity and solve interesting historical questions like where modern humans evolved and how they spread across the globe. This book is a compilation of what he and many researchers have found.

Chambers, Donald A. *DNA: The Double Helix: Perspective and Prospective at Forty Years.* New York: New York Academy of Sciences, 1995. A collection of historical papers from major figures involved in the discovery of DNA with reminiscences from some of the authors.

Chimpanzee Sequencing and Analysis Consortium, The. "Initial sequence of the chimpanzee genome and comparison with the human genome." *Nature* 437 (2005): 69–87. This article presents an in-depth contrast of the complete DNA sequences of humans and chimpanzees.

Crichton, Michael. *The Andromeda Strain.* New York: Knopf, 1969. Crichton's breakthrough novel about the arrival of an alien microorganism on Earth and how scientists struggle to control it.

———. *Jurassic Park.* New York, Knopf: 1990. A best seller about a group of researchers who have obtained DNA from dinosaurs and brought them back to life by reconstructing their genomes with patches from other organisms.

———. *State of Fear.* New York: HarperCollins Publishers, 2004. Crichton's techno-thriller about a group of ecoterrorists who perpetrate global disasters in order to convince the public of the seriousness of global warming and other environmental threats.

Crick, Francis. *What Mad Pursuit: A Personal View of Scientific Discovery.* New York: Basic Books, 1988. Crick's account of dead ends, setbacks, wild ideas, and finally glory on the road to the discovery of the structure of DNA, with speculations on the future of neurobiology and other fields.

Darwin, Charles. *The Descent of Man.* Amherst, N.Y.: Prometheus, 1998. In this book, originally published 12 years after *On the Origin of Species,* Darwin outlines his ideas on the place of human beings in evolutionary theory.

———. *On the Origin of Species.* Edison, N.J.: Castle Books, 2004. Darwin's first, enormous work on evolution, which examines a huge number of facts while building a case for heredity, variation, and natural selection as the forces that produce new species from existing ones.

———. *The Voyage of the* Beagle. London: Penguin Books, 1989. A scientific adventure story; Darwin's account of his five years as a young naturalist aboard the *Beagle.* He had not yet discovered the principles of evolution but was aware of the need for a scientific theory of life. Readers watch over his shoulder as he tries to make sense of questions that puzzled scientists everywhere in the mid-19th century.

Diamond, Jared. *Guns, Germs, and Steel.* New York: W. W. Norton, 2005. A new look at how Western civilization came to dominate the globe, integrating information from archeology, anthropology, genetics, evolutionary biology and many other sources. Many consider Diamond the first to accurately apply evolutionary principles to the question of why some societies become dominant over others.

Dorus, Steve, Eric J. Vallender, et al. "Accelerated evolution of nervous system genes in the origin of *Homo sapiens.*" *Cell* 119, no. 7 (December 29, 2004): 1,027–1,040. A systematic study of how quickly genes crucial to brain evolution have evolved in primates and mammals. The study compares genes in humans, macaque monkeys, mice, and rats, revealing that brain-related genes have been subject to pressure from natural selection, especially in the branch of primates leading to humans.

Dubrova, Yuri, Valeri Nesterov, et al. "Further evidence for elevated human minisatellite mutation rate in Belarus eight years after the Chernobyl accident." *Mutation Research* 381 (2007): 267–278. A groundbreaking study examining the long-term impact of the Chernobyl accident on the genomes of people living in the region.

Elliott, William H., and Daphne C. Elliott. *Biochemistry and Molecular Biology.* New York: Oxford University Press, 1997. An excellent college-level overview of the biochemistry of the cell.

Fruton, Joseph. *Proteins, Enzymes, Genes: The Interplay of Chemistry and Biology.* New Haven, Conn.: Yale University Press, 1999. A very detailed historical account of the lives and work of the chemists, physicists, and biologists who worked out the major functions of the molecules of life.

Fukuyama, Francis. *Our Posthuman Future: Consequences of the Biotechnology Revolution.* London: Profile Books Ltd., 2003. A book by a leading political economist about the economic and social impact of current developments in genetic engineering and other forms of biotechnology, with keen insights into the ethical dilemmas that they pose.

———. *The End of History and the Last Man.* New York: Maxwell Macmillan International, 1992. A philosophical perspective on the development of societies and forms of government, proposing that Western liberal democracy is the logical end point of human social evolution.

Gilbert, Scott. *Developmental Biology.* Sunderland, Mass.: Sinauer Associates, 1997. An excellent college-level text on all aspects of developmental biology.

Goldsmith, Timothy H., and William F. Zimmermann. *Biology, Evolution, and Human Nature.* New York: Wiley, 2001. Life from the level of genes to human biology and behavior.

Gregory, T. Ryan, ed. *The Evolution of the Genome.* Boston: Elsevier Academic Press, 2005. An advanced-level book presenting the major themes of evolution in the age of genomes, written by leading researchers for graduate students and scientists.

Harper, Peter S. *Practical Genetic Counselling,* 6th ed. London: Hodder Arnold, 2004. An introduction to the basics of genetic counseling, with a review of genetic disorders based on body systems such as the nervous system, the eye, and cardiovascular and respiratory diseases, and a section on privacy issues and other themes under the heading "Genetics and Society."

Henig, Robin Marantz. *A Monk and Two Peas.* London: Weiden-feld & Nicolson, 2000. A popular, easy-to-read account of Gregor Mendel's work and its impact on later science.

Huxley, Aldous. *Brave New World.* New York: Perrenial, 1998. Huxley's anti-utopian novel of the future.

Huxley, J. S. *Man in the Modern World.* London: Chatto & Windus, 1947. Originally published in *The Uniqueness of Man,* 1941, the essay by biologist Julian Huxley, brother of the author of *Brave New World,* supports eugenics movements to improve the human race.

Judson, Horace Freeland. *The Eighth Day of Creation: Makers of the Revolution in Biology.* New York: Simon and Schuster, 1979. A comprehensive history of the science and people behind the creation of molecular biology, from the early 20th century to the 1970s, based on hundreds of hours of interviews Judson conducted with the researchers who created this field.

Koch, Christof. *The Quest for Consciousness: A Neurobiological Approach.* Englewood, Colo.: Roberts & Company, 2004. Koch is trying to establish the biological basis of consciousness and related mental abilities in humans. This fascinating book explores discoveries from the activity of genes to the behavior of modules of the brain and presents scientists' best current knowledge of the relationship between the physical brain and the metaphysical mind.

Kohler, Robert E. *Lords of the Fly: Drosophila Genetics and the Experimental Life.* Chicago: University of Chicago Press, 1994. The story of Thomas Hunt Morgan and his disciples, whose discoveries regarding fruit fly genes dominated genetics in the first half of the 20th century.

Lu, Shi-Jiang, Qiang Feng, et al. "Biological properties and enucleation of red blood cells from human embryonic stem cells." *Blood* (prepublished online August 19, 2008). Describes the successful creation of differentiated blood from embryonic stem cells in cell cultures in the laboratory.

Lutz, Peter L. *The Rise of Experimental Biology: An Illustrated History.* Totowa, N.J.: Human Press, 2002. A very readable, wonderfully illustrated book tracing the history of biology from ancient times to the modern era.

Maddox, Brenda. *Rosalind Franklin: The Dark Lady of DNA.* London: HarperCollins Publishers, 2002. An account of the life and work of Rosalind Franklin, who played a key role in the discovery of DNA's structure but who had trouble fitting in to the scientific culture of London in the 1950s.

Magner, Lois N. *A History of the Life Sciences.* New York: M. Dekker, 1979. An excellent, wide-ranging book on the development of ideas about life from ancient times to the dawn of genetic engineering.

McElheny, Victor K. *Watson and DNA: Making a Scientific Revolution.* Cambridge, Mass.: Perseus, 2003. A retrospective on the work and life of the extraordinary scientific personality James Watson, codiscoverer of the structure of DNA.

Musaro, Antonió, Nadia Rosenthal, et al. "Stem cell-mediated muscle regeneration is enhanced by local isoform of insulin-like growth factor 1." *Proceedings of the National Academy of Sciences* (February 3, 2004): 1,206–1,210. A paper by Nadia Rosenthal's group on the effects of a mutation in the IGF-1 gene that stimulates the regeneration of muscle in mice and extends animals' lifespans.

Plotkin, Mark J. *Tales of a Shaman's Apprentice.* New York: Penguin Books, 1993. A landmark book by a pioneer in the field of ethnobotany. For decades, Plotkin has been working with remote tribes on conservation issues and to record their knowledge of pharmacological and other uses of indigenous plants.

Ptashne, Mark, and Alexander Gann. *Genes and Signals.* Cold Spring Harbor, N.Y.: Cold Spring Harbor Laboratory Press, 2002. A readable and nicely illustrated book presenting a modern view of how genes in bacteria are regulated and what these findings mean for the study of other organisms.

Purves, William K., David Sadava, et al. *Life: The Science of Biology.* Kenndallville, Ind.: Sinauer Associates and W. H. Freeman, 2003. A comprehensive overview of themes from the life sciences. The book is most suited for beginning university students, but most of the chapters will be accessible to younger students and teachers. The illustrations used to demonstrate methods in biology and processes such as embryonic development are very clear and informative.

Sacks, Oliver. *The Island of the Colour-Blind and Cycad Island.* London: Picador, 1996. Neurobiologist Sacks's personal account of his travels to the Pacific islands of Cycad and Pingelap. There he encountered people with an unusual genetic condition that allows them only to see shades of gray; the book contains his reflections on the impact of this condition on island culture.

Scott, Christopher. *Stem Cell Now: From the Experiment That Shook the World to the New Politics of Life.* New York: Pi Press, 2006. An excellent, very readable introduction to stem cells and the role that they are likely to play in the medicine of the future, taking into account political, social, and ethical dimensions of their use.

Stent, Gunther. *Molecular Genetics: An Introductory Narrative.* San Francisco: W. H. Freeman, 1971. A classic book for college-level students about the development of genetics and molecular biology by a researcher and teacher who witnessed it firsthand.

Strachan, Tom, and Andrew P. Read. *Human Molecular Genetics 3.* New York: Garland Publishing, 2004. An excellent college-level textbook giving a comprehensive overview of methods and findings in human genetics in the molecular age.

Tanford, Charles, and Jacqueline Reynolds. *Nature's Robots: A History of Proteins.* New York: Oxford University Press, 2001. A history of biochemical and physical studies of proteins and their functions and the major researchers in the field.

Thorson, James. *Aging in a Changing Society,* 2nd ed. Philadelphia: Brunner/Mazel, 2000. Thorson is professor of gerontology at the University of Nebraska. This book presents facts, trends, and a fascinating social perspective on the rising life expectancy in the modern world.

Tudge, Colin. *In Mendel's Footnotes.* London: Vintage, 2002. An excellent review of ideas and discoveries in genetics from Mendel's day to the 21st century.

————. *The Variety of Life: A Survey and a Celebration of All the Creatures That Have Ever Lived.* New York: Oxford University Press, 2000. A beautifully illustrated tree of life classifying and describing the spectrum of life on Earth.

Vogel, Friedrich, and Arno Motulsky. *Human Genetics,* 3rd ed. New York: Springer-Verlag, 1997. A college-level, in-depth overview of human genetics in the molecular age.

Wang, Eric, Greg Kodama, et al. "Global Landscape of Recent Inferred Darwinian Selection for *Homo sapiens.*" *PNAS* 103, no. 1 (2006): 135–140. A study comparing DNA sequences from humans, apes, and rodents. The work reveals human genes that have been subject to natural selection, particularly genes related to brain development and function.

Watson, James D. *The Double Helix.* New York: Atheneum, 1968. Watson's personal account of the discovery of the structure of DNA.

Watson, James D., and Francis Crick. "A Structure for Deoxyribose Nucleic Acid." *Nature* 171 (1953): 737–738. The original article in which Watson and Crick described the structure of DNA and its implications for genetics and evolution.

Wilson, Edward O. *The Future of Life.* London, Abacus: 2002. An easy-to-read overview of the current state of biodiversity throughout the world, with fascinating insights into the impact of human activity on the environment and proposals for coping with overpopulation, species extinctions, and other ecological problems, written by the world's foremost expert on ants and a two-time Pulitzer Prize winner.

Web Sites

There are tens of thousands of Web sites devoted to the topics of molecular biology, genetics, evolution, and the other themes of this book. The selection below provides original articles,

188 ● THE FUTURE OF GENETICS

teaching materials, multimedia resources, and links to hundreds
of other excellent sites.

American Society of Naturalists. "Evolution, Science, and Society: Evolutionary Biology and the National Research Agen-
da." Available online. URL: http://www.rci.rutgers.edu/
~ecolevol/fulldoc.pdf. Accessed April 28, 2009. A document
from the American Society of Naturalists and several other
organizations, summarizing evolutionary theory and showing how it has contributed to other fields including health,
agriculture, and the environmental sciences.

Bradshaw Foundation. "Journey of Mankind—the Peopling of
the World." Available online. URL: http://www.bradshaw
foundation.com/journey/. Accessed April 28, 2009. An online lecture and film giving an excellent visual demonstration
of how and when modern humans likely spread from Africa
to populate the globe.

British Broadcasting Corporation. "BBC—Press Office—Richard
Dimbledy Lecture 2007: Dr. J. Craig Venter." Available online. URL: http://www.bbc.co.uk/pressoffice/pressreleases/
stories/2007/12_december/05/dimbleby.shtml. Accessed April
28, 2009. An interesting lecture from Craig Venter. Venter's was the first individual's genome to be completely sequenced. In the lecture, he describes some anecdotal discoveries about his own genetic code, as an illustration of what
can be learned through personal genomics and with a view
toward a day of personalized medicine.

California Institute of Technology. "The Caltech Institute Archives." Available online. URL: http://archives.caltech.edu/
index.cfm. Accessed April 28, 2009. This site hosts materials
tracing the history of one of America's most important scientific institutes since 1891. One highlight is a huge collection of oral histories with firsthand accounts of some of the
leading figures who have been at Caltech, including George
Beadle, Max Delbrück, and others.

———. "Videos—CNS 120." Available online. URL: http://
www.klab.caltech.edu/cns120/wiki/Videos. Accessed April

28, 2009. A lecture series by Christof Koch, who is carrying out research into the neurobiology of consciousness at Caltech. The series, which can be watched online, gives an excellent overview of Koch's work and the way he has approached questions about the evolution and biology of the human mind.

Center for Genetics and Society. "CGS: Detailed Survey Results." Available online. URL: http://www.geneticsand society.org/article.php?id=404. Accessed April 28, 2009. This article presents the results of numerous surveys conducted in the United States and elsewhere on topics related to genetics, human cloning, and stem cell research, providing a fascinating view of people's knowledge of basic genetic topics as well as how opinions have changed over the past few years.

Department of Energy, Human Genome Project. "Genetics Legislation." Available online.. URL: http://www.ornl.gov/sci/ techresources/Human_Genome/elsi/legislat.shtml. Accessed April 28, 2009. This page presents an overview of legislation regarding human genome information and the protection of personal genetic information. The Web site is a good starting point for teachers and students who want to get an overview of scientific and ethical issues related to human genetics, including information about laws pertaining to genetic testing, patient rights, medical discoveries, etc.

————. "Genetic Disease Information." Available online. URL: http://www.ornl.gov/sci/techresources/Human_Genome/ medicine/assist.shtml. Accessed April 28, 2009. A basic, easy-to-understand guide to the facts about known genetic diseases and ethical and legal issues surrounding diagnoses. The site, which was created by the Human Genome Project, provides links to places where the tests are available and information about new approaches to dealing with the diseases.

Dolan DNA Learning Center, Cold Spring Harbor Laboratory. "DNA Interactive." Available online. URL: http://www.dnai. org. Accessed April 28, 2009. A growing collection of multimedia and archival materials including several hours of filmed interviews with leading figures in molecular biology, a timeline of discoveries, an archive on the American eugenics movement, and a wealth of teaching materials on the topics of this book.

———. "Genes to Cognition Online." Available online. URL: http://www.g2conline.org/. Accessed April 28, 2009. This Web site for students, teachers, and the general public (as well as scientists) offers a huge amount of material on the relationship between genes and thinking and a wide range of related topics. A unique feature of the site is a new, dynamic, style of navigation based on "concept mapping," a learner-directed technique for structuring and visualizing information. The DNA Learning Center is currently testing the site in classrooms to explore new ways of teaching and learning about science.

European Bioinformatics Institute (EBI). "2can." Available online. URL: http://www.ebi.ac.uk/2can/home.html. Accessed April 28, 2009. An educational site from the EBI—one of the world's major Internet providers of information about genomes, proteins, molecular structures, and other types of biological data. Many of the tutorials and basic introductions to the themes are accessible to pupils or people with a bit of basic knowledge in biology.

Exploratorium. "Microscope Imaging Station." Available online. URL: http://www.exploratorium.edu/imaging_station/index. php. Accessed April 28, 2009. San Francisco's Exploratorium is an interactive science museum; its Web site has a range of wonderful activities based on biological themes such as development, blood, stem cells, and the brain. There are also videos, desktop wallpapers that can be downloaded for free, and feature articles on current themes from science.

Institute of Human Origins. "Becoming Human: Paleoanthropology, Evolution and Human Origins." Available online. URL: http://www.becominghuman.org/. Accessed April 28, 2009. An attractive site with a focus on paleoanthropology and human origins, with a video documentary that can be watched online or downloaded, classroom resources, and articles on

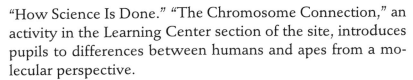

"How Science Is Done." "The Chromosome Connection," an activity in the Learning Center section of the site, introduces pupils to differences between humans and apes from a molecular perspective.

National Academies Press. "The New Science of Metagenomics: Revealing the Secrets of Our Microbial Planet." Available online. URL: http://books.nap.edu/catalog.php?record_id=11902. Accessed April 28, 2009. A fascinating report on biodiversity, summarizing what has been learned by metagenomics projects to sequence the DNA of microbes that inhabit the human body and various ecospheres. The report provides perspectives for future projects and identifies eight key areas in which metagenomics may be useful.

National Center for Biotechnology Information. "Bookshelf." Available online. URL: http://www.ncbi.nlm.nih.gov/sites/entrz?db=books. Accessed April 28, 2009. A collection of excellent online books ranging from biochemistry and molecular biology to health topics. Most of the works are quite technical, but many include very accessible introductions to the topics. Some highlights are: *Molecular Biology of the Cell, Molecular Cell Biology,* and the *Wormbook.* There are also annual reports on health in the United States from the Centers for Disease Control and Prevention.

National Geographic. "Outpost: Human Origins @ nationalgeographic.com." Available online. URL: http://www.nationalgeographic.com/features/outpost/. Accessed April 28, 2009. A virtual expedition, accompanying human fossil hunter Lee Berger on a search for ancient human remains in Botswana and South Africa.

National Health Museum. "Access Excellence: Genetics Links." Available online. URL: http://www.accessexcellence.org/RC/genetics.php. Accessed April 28, 2009. Links and resources from the "Access Excellence" project of the National Health Museum.

National Public Radio. "Wild Cows Cloned." Available online. URL: http://www.npr.org/templates/story/story.php?storyId=1225049. Accessed April 28, 2009. An audio interview

with researcher Robert Lanza, pioneering stem cell researcher (see chapter 1). In this interview Lanza discusses his cloning of an endangered wild cow called the banteng, using cells
taken from an animal that had died 23 years earlier. Lanza
also appears in an interview from 2006 on more general
themes in stem cell research at the following URL: http://
www.npr.org/templates/story/story.php?storyId=5204335.
The NPR Web site offers a wide range of interviews and
broadcasts on biological and medical themes that can be lis-
tened to online.

Nobel Foundation. "Video Interviews with Nobel Laureates
in Physiology or Medicine." Available online. URL: http://
nobelprize.org/nobel_prizes/medicine/video_interviews.
html. Accessed April 28, 2009. Video interviews with laureates from the past four decades, many of whom have been
molecular biologists or researchers from related fields. Follow links to interviews with winners of other prizes, Nobel
lectures, and other resources.

Patricia Piccinini. "Speculative Fabulations for Technoculture's
Generations: Taking Care of Unexpected Country," by Donna Haraway. Available online. URL: http://www.patriciapic
cinini.net/. Accessed April 28, 2009. This article is a critical
review of several years of Piccinini's sculptures, and can be
found under the essays link on the artist's Web site. Piccinini
imagines what genetic hybrids of humans and other animals
might look like in the future and creates very real-looking
representations of them.

Public Broadcasting Service (PBS). "American Experience: Jesse
James." Available online. URL: http://www.pbs.org/wgbh/
amex/james/index.html. Accessed April 28, 2009. This is the
home page of a PBS documentary centered on the life of Jesse
James, including the transcript of the broadcast, a wide range
of images, and other supplementary materials.

Research Collaboratory for Structural Bioinformatics. "RCSB
Protein Data Bank." Available online. URL: http://www.rcsb.
org. Accessed April 28, 2009. This site provides "a variety of
tools and resources for studying the structures of biological

macromolecules and their relationships to sequence, function, and disease." There is a multimedia tutorial on how to use the tools and databases. One special feature is the "Molecule of the Month," with beautiful illustrations by David Goodsell.

Science Friday. "Science Friday Archives: Cancer Update with RobertWeinberg." Available online. URL: http://www.science friday.com/program/archives/200710123. Accessed April 28, 2009. Science Friday is heard live every Friday on National Public Radio stations. This link points to the podcast of a program originally broadcast on Oct. 12, 2007, with Robert Weinberg, renowned cancer researcher at MIT. Weinberg's group had just discovered that small molecules called microRNAs regulate the production of proteins in tumor cells, a finding with significant implications for diagnosis and therapy.

Scientific American. "The First Human Cloned Embryo." Available online. URL: http://www.sciam.com/article.cfm?id=the-first-human-cloned-em. Accessed April 28, 2009. An article by Robert Lanza and his colleagues describing the methods they used to make the first clones of human embryonic stem cells. The article explains the methods used and the researchers' hopes for how this type of work may change medicine in the future.

TalkOrigins. "The Talk Origins Archive." Available online. URL: http://www.talkdesign.org. Accessed April 28, 2009. A Web site devoted to "assessing the claims of the Intelligent Design movement from the perspective of mainstream science; addressing the wider political, cultural, philosophical, moral, religious, and educational issues that have inspired the ID movement; and providing an archive of materials that critically examine the scientific claims of the ID movement." (A subsection of the site deals specifically with the origins of humans: http://www.talkorigins.org/faqs/homs.)

Tech Museum of Innovation, San Jose, California. "Understanding Genetics: Human Health and the Genome." Available online. URL: http://www.thetech.org/genetics. Accessed April

28, 2009. An excellent collection of news and feature stories on scientific discoveries and ethical issues surrounding genetics.

University of California, San Francisco. "Live Long and Prosper: A Conversation About Aging with Cynthia Kenyon—Science Café—UCSF." Available online. URL: http://www.ucsf.edu/science-cafe/conversations/kenyon. Accessed April 28, 2009. UCSF's Science Café features interviews with university scientists about their work—at a level easy to understand for the general public. This link is devoted to the research of Cynthia Kenyon, director of the Larry L. Hillblom Center for the Biology of Aging and a pioneer in the biology of aging in the worm *C. elegans* and other model organisms. Visitors to the Web site can read the article or listen to it in mp3 format. Kenyon gives an overview of her own work for nonscientists at the home page of her own lab: http://kenyonlab.ucsf.edu/html/non-scientist_overview.html.

University of California, Santa Cruz. "UCSC Genome Bioinformatics." Available online. URL: http://genome.ucsc.edu/bestlinks.html. Accessed April 28, 2009. A portal to high-quality resources for the study of molecules and genomes, from UCSC and other sources. The map of the BRC2A gene presented in chapter 2 was obtained using this site.

University of Cambridge. "The Complete Works of Charles Darwin Online." Available online. URL: http://darwin-online.org.uk. Accessed April 28, 2009. An online version of Darwin's complete publications, 20,000 private papers, and hundreds of supplementary works.

University of Utah, Genetic Science Learning Center. "Learn Genetics." Available online. URL: http://learn.genetics.utah.edu/. Accessed April 28, 2009. An excellent Web site introducing the basics of genetic science, including a "Biotechniques Virtual Laboratory," special features on the genetics and neurobiology of addiction, stem cells, and molecular genealogy, and podcasts on the genetics of perception and aging.

University of Washington. "Gene Tests Home Page." Available online. URL: http://www.genetests.org/. Accessed April 28, 2009. A site mainly aimed at professionals interested in obtaining up-to-date information on new links between genes and disease, with links to articles on the latest research, laboratories carrying out studies of particular diseases, and ongoing clinical trials. But the site also has a wealth of information on the genetic tests that are available, links to doctors and clinics who provide specific tests, a lexicon of technical terms, and educational resources.

University of Washington Television. "UWTV Program: Genomic Views of Human History." Available online. URL: http://www.uwtv.org/programs/displayevent.aspx?rID=2493. Accessed April 28, 2009. A lecture from Mary-Claire King (see chapter 4) that can be watched online. The theme is how new tools of genomic analysis are being used to investigate the genes of modern humans, shedding light on historical puzzles such as ancient migrations and the settlement of the globe.

Vega Science Trust. "Scientists at Vega". Available online. URL: http://www.vega.org.uk/video/internal/15. Accessed April 28, 2009. Filmed interviews with some of the great figures in 20th-century and current science, including Max Perutz, Kurt Wüthrich, Aaron Klug, Fred Sanger, John Sulston, Bert Sakmann, Christiane Nüsslein-Volhard, etc.

Wisconsin Medical Society. "Wisconsin Medical Society." Available online. URL: http://www.wisconsinmedicalsociety.org/savant_syndrome. Accessed April 28, 2009. A site devoted to the work of Darold Treffert, a psychiatrist who is likely the world's foremost expert on people with savant syndrome. It includes an overview of the field, references to the literature, and written and video portraits of over 20 savants.

Index

Italic page numbers indicate illustrations or maps.

A

adaptation 18
adenine (A) 28, 32
ADHD (attention deficit/hyperactivity disorder) 126–127
Aequorea victoria 110
aequorin 111
Affymetrix 93, 95
aging, counteracting 140, 145–151, *151*
Aging in a Changing Society (Thorson) 134
alkaloids 137–138
alleles 13, 18, 89
Allen, Myles 86
Alzheimer, Alois 131
Alzheimer's disease 131–132, *133*, 135, 138
Amazon (film) 136
Amazon rain forests 136–137, *137*
American Breeders Association 65
amino acids 48, 101
amyloid plaques 131–132, 134
The Andromeda Strain (Crichton) 83–84
anthrax 7–8, 10

antibiotic resistance 10, 74
antibodies 95, 139
Antinori, Severino 71
ants 152, 154
apoptosis 105
APP (amyloid precursor protein) 131–132, *133*, 135
Arber, Werner 42
Arendt, Detlev 47
arrays. *See* DNA microarrays; tiling arrays
Asimov, Isaac 123
Astbury, William 29–31
attention deficit/hyperactivity disorder (ADHD) 126–127
Australia 70
Avery, Oswald 27–28
axolotl 39

B

Bacillus thuringiensis 77
bacteria
    Avery's research 27–28
    for bioremediation 43–44
    for conditional mutagenesis 106
    and gene sequencing 46
    and genetic engineering 43

    Griffith's genetic studies 27
    Koch's research 7–8
bacteriorhodopsin 99
Baer, Karl Ernst von 35
bases. *See* nucleotides
Bateson, William 14–16
B cells 139
The Beginning and the End (Asimov) 123
Benedict XVI (pope) 62
Benin 156
the Bible 59, 61, 62
bigfoot 88–89
biodiversity 97–104
biofilms 43
bioinformatics 47
bioremediation 43–44
biotechnology. *See* genetic engineering
Bismarck, Otto von 6
bison 89
Blackburn, Elizabeth 146
Bloch, Felix 119
blood vessels 121–122
bollworm 81, 82
Bork, Peer 99
Boveri, Theodore 26
bovine spongiform encephalopathy 77
Boyer, Herbert 43
Bragg, Sir Lawrence 31
brain 121–122, 140–145
brain diseases 129–135, 138–140